T0258640

Geometrical Landscapes

WRITING SCIENCE

EDITORS   Timothy Lenoir and Hans Ulrich Gumbrecht

Geometrical Landscapes

THE VOYAGES OF DISCOVERY AND THE
TRANSFORMATION OF MATHEMATICAL PRACTICE

Amir R. Alexander

STANFORD UNIVERSITY PRESS
STANFORD, CALIFORNIA

Stanford University Press
Stanford, California

© 2002 by the Board of Trustees of the
Leland Stanford Junior University

Printed in the United States of America
on acid-free, archival-quality paper

ISBN 0-8047-3260-4 (cloth : alk. paper)

Original Printing 2002
Last figure below indicates year of this printing:
11   10   09   08   07   06   05   04   03

Typeset by BookMatters in 10/13 Sabon

To my father
Shlomo Alexander
1930–1998

CONTENTS

ILLUSTRATIONS

ACKNOWLEDGMENTS

This project has been my companion for many years now, and I have talked about it with practically anyone who was willing to listen. Although I would like to thank here all those patient listeners, the list that follows is unavoidably incomplete.

The seeds of this work were sown when I was a first-year college student in Jerusalem. I enrolled in a course on the Italian Renaissance offered by Rivka Feldhay simply because it fit my schedule. What I learned in that freshman seminar about cultural history and the history of science has altered the course of my intellectual life. During my years in Jerusalem my interest continued to develop with the help of Yehuda Elkana, who supported me intellectually and otherwise at the Van Leer Institute in Jerusalem, and Michael Heyd, my advisor at the Hebrew University. Gabi Herman, a friend as well as a teacher, shared his academic experiences and encouraged me to pursue the study of history as a vocation. Many thanks to them all.

The bulk of the research for this book was done during my graduate studies at Stanford. My friends in graduate school endured many hours of conversation on the cultural history of mathematics and helped sustain me through those hard but enormously stimulating years. My teachers at Stanford deserve special thanks. Keith Baker taught me the importance of language and rhetoric in a graduate seminar on the French Revolution. Peter Galison showed me how to straddle the divide between the history and the philosophy of science. Hans Gumbrecht introduced me to a new world of literary studies and helped me use these approaches in my own work. Many thanks as well to Paula Findlen, Lou Roberts, and Paul Seaver. I owe a great debt to the late Wilbur Knorr, a fellow historian of mathematics, and share his friends' and colleagues' sense of loss because of his sudden passing.

Many stimulating discussions with Mario Biagioli over the years helped make this book what it is. Amos Funkenstein challenged and provoked me in many conversations, sadly cut short by his untimely death. Simon Schaffer and Moti Feingold read my entire dissertation. Their many insightful comments from their different perspectives forced me to rethink some fundamental assumptions and helped make this a better book. Peter Dear read parts of this manuscript and used his vast knowledge of early modern European intellectual history to provide scholarly and illuminating comments. Michael Mahoney made very helpful suggestions from a historian of mathematics' perspective. He also directed me to the passage that is the epigraph of this book. Joan Richards read the entire manuscript, and her detailed comments made this a much better book. Nathan MacBrien and Sumathi Raghavan of Stanford University Press saw the manuscript through the acquisition stage. Dr. Silvia Busch translated from the Italian the passages cited in Chapter 6; Laura Moss Gottlieb produced the index; and Anna Eberhard Friedlander, my production editor at Stanford University Press, oversaw the process of turning a manuscript into a book. Many thanks to them all.

Many thanks to my friends and colleagues at the University of California, Los Angeles. Margaret Jacob read the entire manuscript and was generous with her scholarly insights, as well as with her time and support. Ted Porter and Peter Reill helped bring me to UCLA and provided me with academic homes at the Center for the Cultural Studies of Science and the Center for 17th and 18th Century Studies. I spent many hours of conversation with Sharon Traweek and Mary Terrall, and I thank them for their friendship and intellectual companionship.

This book was written with the help of numerous fellowships and grants. A four-year Stanford University graduate fellowship supported me during my early years in this country and made this project possible. Travel awards from the Stanford Center for European Studies and the Stanford Institute for International Studies supported me during a crucial research trip to England. A Mabel McLeod Lewis grant and a Weter dissertation completion grant helped support me during my final two years in graduate school. A fellowship at the Stanford Humanities Center in 1993–94 provided me with intellectual companionship and ideal writing conditions. A two-year Mellon postdoctoral fellowship at the UCLA Humanities Consortium gave me the time and opportunity to turn a doctoral dissertation into a book.

With the exception of Chapter 6, all of this book's text appears here for the first time. A version of Chapter 6 was published as "Exploration Mathematics: The Rhetoric of Discovery and the Rise of Infinitesimal Methods" in the spring 2001 issue of *Configurations*. A brief and early version of the argument regarding Thomas Hariot appeared as "The Imperialist

Space of Elizabethan Mathematics" in volume 26 (1995) of *Studies in the History and Philosophy of Science*.

The influence of Timothy Lenoir pervades this book. He was with this project from beginning to end, and I can honestly say that without him this book would not have been written. My wife Bonnie has been with me on this entire journey, from my search for a dissertation topic to the publication of this book. My thanks, and my love, go out to her.

My father, Shlomo Alexander, was my first and most important intellectual mentor. He taught me to love both history and mathematics and started me on the course that has led to this book. He died in a traffic accident in the summer of 1998. I dedicate this book to him.

Hac tempestate in geometricis inventum et superatum feliciter esse Bonae Spei promontorium illud, unde expedita existat navigatio ad inaccessas ante tetragonismorum praesertim regions.

At this time in geometry, the Cape of Good Hope was fortunately discovered and circumnavigated; whence navigation to regions practically inaccessible before the quadratures was made possible.

ANTOINE DE LALOUVERE,
Veterum geometria promota (Toulouse, 1660)

Introduction

In a famous passage of the *New Organon*, Sir Francis Bacon challenged the natural philosophers of his time to live up to the example of geographical explorers. "It would be disgraceful," he wrote, "if, while the regions of the material globe—that is, of the earth, of the sea, of the stars—have been in our times laid widely open and revealed, the intellectual globe should remain shut out within the narrow limits of old discoveries." Bacon was not alone in this view: the great voyages of exploration were repeatedly cited as a model and an inspiration by early modern promoters of the new sciences. The image of the natural philosopher as a Columbus or Magellan, pushing forward the frontiers of knowledge, became a commonplace of scientific treatises and pamphlets of the period. The newly discovered lands and continents seemed both a proof of the inadequacy of the traditional canon and a promise of great troves of knowledge waiting to be unveiled.[1]

The rhetoric of exploration, however, did not promote all forms of knowledge equally: taking its cue from the explorers themselves, it emphasized direct observation and experience over abstract reasoning. It was, after all, only through actual travel and personal experience that the great voyagers discovered new lands and oceans, forever undermining the credibility of the traditional scholars who based their geographical speculations on the ancient sources. If the new sciences are to follow the lead of the explorers, they should be based on actual "experience" and not on disembodied reason or speculation.

It is hardly surprising that Bacon and fellow promoters of the experimental and observational approach were quick to adopt the imagery of geographical discovery. The heroic explorer, searching for hidden riches in an unfamiliar land, was a perfect role model for a natural philosopher seeking to unveil the hidden secrets of nature.[2] Mathematicians, however, seemed to be left out of the fold. Mathematics was viewed as pure and abstract, the epitome of disembodied reasoning. It did not explore the world in search of new

1

knowledge or seek to discover new phenomena. The truth of its propositions was not based on induction from empirical observation, but instead on unfaltering deduction from general postulates. Rather than interrogate nature in search of her hidden secrets, mathematics posited universal laws of quantity and magnitude that nature must follow. Such a field would indeed have little use for the imagery of exploration.[3]

Significantly, however, some of the most important developments of early modern mathematics radically deviated from this traditional approach. In particular, infinitesimal methods and the early forms of the calculus went against the grain of accepted mathematical practices. Whereas traditional practitioners sought to deduce necessary truths from universally accepted assumptions, mathematicians such as Bonaventura Cavalieri, Evangelista Torricelli, Thomas Hariot, and John Wallis sought to investigate a given object—the mathematical continuum. Instead of proceeding through rigorous deduction, the pioneers of infinitesimals adopted a more intuitive non-rigorous approach to their craft and sometimes relied on mere trial and error.[4] Whereas mathematics' claim to truth had always been based on its irrefutable logic, the study of infinitesimals involved confronting well-established paradoxes and contradictions. The pioneers of infinitesimals, in other words, were willing to stretch mathematical logic to the limit in order to explore the secrets of the mathematical continuum. Theirs was indeed a new kind of mathematics.[5]

How did such a shift in the aims and practices of mathematics come about? It is the main contention of this book that this shift was made possible by a new vision of mathematics that gained acceptance during this time. In this view, mathematics was an adventurous journey, a voyage of exploration and discovery in search of hidden marvels and gems. Despite its apparent incompatibility with their subject matter, some early modern mathematicians adopted the imagery of geographical discovery and made it their own. Identifying with the celebrated explorers, they came to view themselves as adventurous voyagers in their own right. Whereas Columbus and Magellan had sailed actual oceans and mapped unknown territories, their mathematical successors would uncover the secrets of quantity and magnitude in the uncharted lands of mathematics. For them, mathematics was no longer concerned with the elaboration of universal truths, but rather with the exploration of unknown objects. It was a novel stance, which thoroughly revised the very definition of the field, and placed the "discovery" of new knowledge ahead of rigorous "proof" of known results. It was a stance, moreover, that was uniquely suited for the study of infinitesimals. Infinitesimal methods, I will argue, were a mathematics of exploration.[6]

The book proceeds chronologically, starting in late medieval times and ending in the late seventeenth century. The first chapter identifies a new kind of rhetoric of exploration that became prevalent in the sixteenth century and was embodied in the image of the courageous voyager, who traveled to the ends of the earth in search of great conquests and unimaginable riches. This imagery, needless to say, both inspired and was itself molded by the conquest of the Americas. Its roots, however, are much older. The chapter demonstrates how the new vision of exploration combined elements from two distinct and preexisting Western traditions—the crusading histories and the chivalric romances. Both traditions contributed certain aspects to the imagery of discovery, but their combination in the form of the narrative of exploration was novel indeed.

Chapter 2 is a study of the ways in which the new rhetoric of exploration shaped the actual exploration and mapping of the New World. It is also the beginning of a special focus on the English experience of exploration in Chapters 2 through 5. The chapter focuses on the pamphlets and treatises generated by several well-known enterprises of exploration: Martin Frobisher's expeditions to the Canadian arctic between 1576 and 1580; Walter Raleigh's attempts to colonize Virginia between 1584 and 1587; Raleigh's search for the lost kingdom of El Dorado in the 1590s; and Foxe's and James's voyages to Hudson's Bay in the 1630s in search of the Northwest Passage. In each of these cases, the chapter will demonstrate how the texts were shaped by the narrative of exploration and discovery discussed in Chapter 1. Furthermore, the chapter argues that not only the texts but also the geography of the land, as presented in the explorers' maps, was molded by the same narrative. The prevalent rhetoric of exploration and discovery tended to produce a standardized "geography" of undiscovered lands. Although this section is not generally concerned with mathematicians, their presence is, nonetheless, inescapable. Treatises and maps by Thomas Hariot and Henry Briggs point to the active role of mathematicians in the promotion and diffusion of the tale of exploration.

Chapter 3 is the first to focus specifically on mathematicians, although some of them already made an appearance in Chapter 2. English mathematicians, it is shown, were active and enthusiastic participants in the exploration enterprise. They not only lent their technical expertise to support the voyages but also promoted and publicized them using the familiar narrative of geographical exploration. Most importantly, the mathematicians adopted this imagery and applied it to their own trade, describing themselves as daring voyagers on the uncharted mathematical oceans.

Chapter 4 centers on Thomas Hariot, who was the most original English mathematician of the age, as well as a geographical explorer in Walter Raleigh's Virginia colony. He was thoroughly versed in the rhetoric of exploration, which structures much of his geographical writings as well as his mapping of the new lands. Hariot, I show, adopted the geographical explorers' epistemology and applied it to his scientific studies. Like the great discoverers of the age (and in contrast to more traditional views), he insisted on the primacy of firsthand observation as the proper way to true knowledge. He then adopted the explorers' account on what was in fact "seen" or "discovered"—namely, the narrative of exploration. Both elements, which Hariot borrowed from the exploration literature, are constitutive of his optical theory of refraction. The passage of a ray of light through a material medium is constructed as voyage of exploration and discovery into the inner recesses of matter. The hidden treasures within, the barriers protecting them, and the open passages eventually discovered, all make their appearance in Hariot's theory of refraction. Hariot's treatise on the kinematics of reflecting balls further elaborates this tale: it is meant as a guide into the wondrous secrets hidden within the structure of matter.

Hariot's mathematics, and in particular his signature mathematical atomism, is the subject of Chapter 5. Hariot was one of the earliest proponents of the use of mathematical infinitesimals, and he used them skillfully to obtain novel and groundbreaking results. A close study of his unpublished manuscripts reveals that Hariot's views on the structure of the continuum originated in his work on a classical nautical problem, namely, producing a chart that would preserve true directions. In working on this problem, he analyzed the structure of a complex curve—the equiangular spiral—by treating it as composed of an infinite number of discrete parts. This led Hariot to a wider consideration of the structure of the continuum in his unpublished treatise "De Infinitis." In this tract Hariot posed a long series of confusing and seemingly insoluble problems concerning the composition of the continuum, which he referred to as a "Labyrinth of Daedalus" and a "Gordian Knot." For Hariot, there was only one solution to this confusion: the seemingly smooth continuum must be penetrated by equally spaced breaks. The apparently impenetrable continuum, according to Hariot, was in fact broken by regularly placed cleavages, which offered access to the great mysteries it concealed. Hariot's mathematical atomism and advocacy of infinitesimals, it is shown, are directly informed by his vision of mathematics as a voyage of discovery.

Chapter 6 moves beyond Hariot's work and traces the role of the rhetoric of exploration in the development of infinitesimal methods in the seventeenth century. Although the heroic age of the voyages was now past, its imagery persisted in mathematical circles. Italian and English mathematicians continued to refer to their field in terms borrowed from geographical exploration. Their mathematical practices were shaped accordingly. Viewing themselves as voyagers in unknown lands, they sought to unveil the secrets of the elusive mathematical continuum and map its contours. Evangelista Torricelli in Italy and John Wallis in England led the way in developing mathematical techniques designed to uncover its hidden laws. They approached mathematics experimentally, often using trial and error in order to detect the symmetries of number and magnitude that structure the continuum.

Whereas classical mathematicians carefully avoided the logical minefield of infinitesimals and the composition of the continuum, the new approach legitimized and encouraged their study. The systematic attempt to analyze the continuum in the face of paradoxes and contradictions fit in well with the legendary tales of exploration and discovery. Like true voyagers, the mathematicians sought to unveil hidden secrets, and like them they were willing to risk shipwreck and ruin. As explorers of dangerous lands, they allowed themselves to proceed with their work even when facing the dangerous paradoxes that had stopped classical practitioners in their tracks. Their ultimate goal was to map the secrets of the continuum and bring them to light. Rigor and precision, the traditional hallmarks of mathematical excellence, were subjected to this goal.

In all these ways the founders of the infinitesimal approach modeled themselves on the heroic tales of the great voyages. The notion of natural philosophy as an exploration of the natural world had served to promote an empirical and experimental approach to the study of nature; in much the same way, the vision of mathematics as a hazardous voyage of discovery supported the development of the new infinitesimal mathematics.

CHAPTER I

Prologue: The Origins of the Exploration Narrative

On the night of December 25, 1492, the *Santa Maria*, flagship of Christopher Columbus, ran aground on a reef off the island of Hispaniola. No men or stores were lost in this mishap, Columbus recorded in his *Diario*, and in fact, the destruction of the ship was a blessing in disguise. The unwieldy vessel was in any case "unsuited for the work of exploration" in his opinion. Her timbers could better be used to build a permanent fort and settlement on the site, and her supplies would serve to stock it until the admiral's return the following year. Columbus christened this first European settlement in the New World Navidad, because it was founded on Christmas Day.[1]

Despite the fact that the town of Navidad was born of an accident, Columbus had high hopes for it, and it figured prominently in his plans. He was confident, he wrote in the *Diario*, "that on the return he would undertake from Castile he would find a barrel of gold that those who were left would have acquired by exchange; and that they would have found the gold mine and the spicery."[2] Surprisingly, however, the projected economic success of the town was not aimed at establishing a secure foothold for the Spanish Crown in the Indies, or even at enriching the home country. Columbus had loftier goals in mind: with the riches acquired in Navidad, he suggested, "the sovereigns, before three years [are over], will undertake and prepare to go conquer the Holy Sepulchre; for thus I urged your highness to spend all the profits of this my enterprise on the conquest of Jerusalem."[3] The town of Navidad, in Columbus's eyes, was the first stage of a crusade to liberate Jerusalem and return the Holy Sepulchre to Christian hands. It was a stepping stone not to the settlement of an uncharted New World, but rather back to Jerusalem, the heart and center of the old one.[4]

That Columbus was serious about his crusade is hardly open to doubt: he returned to the subject again and again throughout his life. In his will of 1498 he explained that when he embarked on his discoveries, "it was with the intention of supplicating the King and Queen, our lords, that whatever moneys should be derived from the said Indies should be invested in the conquest of Jerusalem."[5] A letter to Pope Alexander VI four years later shows that he had also given careful consideration to the practical challenges of his proposed expedition: in order to liberate Jerusalem, he hopefully informed the Pope, he would need one hundred thousand foot soldiers and ten thousand horses under his command.[6]

While the economic and military aspects of the project are mentioned in Columbus's correspondence, the spiritual connections between the crusade to liberate Jerusalem and Columbus's enterprise of the Indies are dealt with in the *Book of Prophecies* of 1498, which Columbus was writing during the last decade of his life.[7] Indeed, the opening passage already sets up the central theme of the *Book*: "Here begins the book, or handbook, of sources, statements, opinions and prophecies on the subject of the recovery of God's Holy City and Mount Zion, and on the discovery and evangelization of the islands of the Indies and of all other peoples and nations."[8] The discovery of the Indies and the liberation of Jerusalem are, for Columbus, parts of the very same enterprise.

Throughout his career, in both words and deeds, Columbus presented himself as a crusader fighting for the glory of God. He viewed his voyage not as a peaceful survey of foreign lands, but as a divinely inspired military expedition. The opening sentence of his letter announcing the discoveries of his first voyage is typical in this respect. "I know that you will be pleased," he wrote his friend Louis Santangel, "at the great victory with which Our Lord has crowned my voyage."[9]

Stephen Greenblatt has already noted that the reader of such an opening would expect an account of the clash of great armies, rather than the relatively peaceful island hopping that actually followed.[10] It seems more befitting of the imaginary crusading host that Columbus proposed to Alexander VI a decade later, than of his actual flotilla of small merchant vessels. But Columbus viewed his voyage as the opening stage of a military campaign to recapture the Holy Land. Armed conflict was undoubtedly in short supply on this first voyage. But hazards from both nature and humans were plentiful, as were the military virtues of courage and perseverance among the admiral and his men. In Columbus's eyes, the voyage was indeed a military campaign for the glory of God.[11]

For Columbus, it seems, the discovery and exploration of the Indies could only be understood and justified as part of a larger eschatological project. Indeed, as early as 1493 he began signing his name "Christo-ferens," bearer of Christ, as a sign that he had been selected by God to carry His word overseas.[12] The enterprise of the Indies was not a goal in itself, but only a stage in the inevitable triumph of Christianity and the recovery of its holy sites. It is ironic that the man who did more than any other to break down the boundaries of medieval geography saw himself as traveling not to the ends of the earth, but rather toward the center of the world.

The conquistadors who followed Columbus to the New World in subsequent years were equally eager to present themselves as crusaders on a divine mission. Cortés, Díaz, and their fellow adventurers presented themselves as bearers of the cross and spreaders of the gospel. Unlike the admiral, they did not view the settlement of the Indies as leading to the recovery of Jerusalem and the Holy Sepulchre. Instead, the conquest and settlement of the New World became for them a goal unto itself, a crusade in its own right.[13] This conceit was not lost on Bartolomé de Las Casas, denouncer of Spanish atrocities in the New World, who complained repeatedly about the false pretensions of his fellow Spaniards in the New World. What he viewed as the massacre of innocents for greed and personal gain, they present as a glorious crusade of conquest and conversion.[14]

The crusading vision had profound consequences for the way in which the New World was settled. It supported the notion that America would be won by force of arms, and it suggested that the local inhabitants be viewed as hostile heathens who must be converted or annihilated. But perhaps the most crucial effect of the equation of the conquest of the New World with a crusade was one so general and widespread as to be almost invisible: it was the conquistadors' uninhibited and palpable desire to possess and settle the newfound lands.

Stephen Greenblatt has pointed out that this desire was by no means self-evident, but rather highly unusual: when Columbus "took possession" of the island of San Salvador in 1492 he was clearly breaking with the precedents set by medieval travelers. It had never occurred to Marco Polo or the reputed "John Mandeville" to claim for their countries any of the lands they traveled in. But Columbus, who had in all likelihood read both accounts, unhesitatingly "took possession" of the first island on which he landed.[15] Although this gesture was completely alien to the conduct of medieval travelers, it was fully in accord with the crusading tradition, both of the Holy Land and of the Spanish Reconquista. The whole point of a crusade, after all, was to physically occupy the sites it targeted, whether in Jerusalem or in Grenada. This was the model for Columbus's actions, and it was adopted by practically all the explorers in the Age of Discovery who followed in his footsteps.

But if the conquest of the Indies was a crusade, it was, undoubtedly, an odd one. A proper crusade was, of course, a campaign to conquer and liberate Jerusalem. Its target was the center of the world, both geographically and spiritually. It aimed to regain for Christianity its own lost origins, to bring Christendom closer to its roots by physically possessing the sites of its birth. It was, in its way, an attempt to reverse the fall of man and to counter the spiritual corruption accumulated during a millennium by returning to the place where it all began.[16] The conquest of the Indies would certainly not qualify as such a crusade, because it aimed to possess lands that were completely bereft of any Christian past. One could not "recover" lands that were never in one's possession.

To be sure, not all crusades that had been launched or advocated since the first one in 1095 aimed at liberating the Holy Land. Many aimed at nothing more than protecting the settled lands of Christendom from external threats. Thus, for example, we have crusades launched repeatedly against the Ottoman Turks in the fifteenth and sixteenth centuries, and in the Baltic regions periodically throughout the late Middle Ages. Closest to the hearts and experiences of the New World conquistadors was, of course, the Spanish Reconquista, officially designated as a crusade by the papacy. Hundreds of years of fighting had succeeded in gradually and systematically reducing the Islamic foothold in Spain, leading finally to the surrender of Grenada in the very year that Columbus set sail.[17]

These "diverted" crusades did not, of course, replicate the original crusade's insistence on recapturing the unadulterated "source" of the faith at the center of the world. They did, however, share its basic ethos and purpose. A campaign to defend Hungary against the Turks, for example, was obviously not as ambitious as an expedition to recapture the Holy Sepulchre in Jerusalem. But even though such an expedition did not seek to regain the pure "origin" of Christianity, it did aim at preventing the further degradation of ancestral Christian lands through heathen occupation. And although it did not bring the faithful closer to the sources of their belief, it aimed to prevent them from drifting even further away. The crusade of the Spanish Reconquista was even closer in spirit to the Jerusalem crusade, because it sought not only to defend Christian lands but also to recover lost grounds through force of arms. Like the heroic knights who captured Jerusalem in 1099, the soldiers of the Reconquista looked back to an ancient age of Christian supremacy and fought to recover what had been lost.[18]

None of this applied to the self-styled crusaders of the New World. The traditional crusade was a conceptually conservative notion, drawing one backward in time and pulling one toward the spiritual and geographical center.[19] In contrast, the conquistadors sailed to the edge of the known world and beyond, conquering uncharted lands and seeking an unfathomable future. Rather than pointing the way to the recovery of lost territories, they invaded lands that were never even heard of by their forebears and settled among peoples whose origin was completely unknown. If the Spanish settlement of the Indies was a crusade at all, then it was an inverted one, exploding toward the corners of the earth, rather than imploding back on itself toward the center. It is hardly a coincidence that, unlike these other campaigns, the conquest of the New World was never officially designated as a "crusade" by the papacy.

Columbus himself never recognized the tensions between his voyages and the traditional crusading expeditions. His insistence that Navidad was merely a stop on the road to Jerusalem, bears witness to his effort to incorporate his new discoveries within the old framework of crusading. His claim on his third voyage to have located the Earthly Paradise in the Indies shows a similar attitude. Like Jerusalem, the Earthly Paradise was a marker of a lost and pure origin, and Columbus's hope to find it was a vestige of the crusader's effort to recover lost spiritual grounds. Indeed, what would be a surer way to reverse man's fall and lead him back to a pure unadulterated source than the rediscovery of paradise?[20] It is typical of Columbus that even when he appeared to observe the New World in itself, what he ultimately discovered were the sources of his own faith.[21]

Such single-minded determination to deny the essential novelty of the Americas, however, did not survive the admiral. Even in his lifetime, Peter Martyr, who wrote one of the earliest "histories" of the New World, ridiculed Columbus's claim to have located the Earthly Paradise. Las Casas, a few decades later, spoke of America as paradise only metaphorically, in order to highlight the destruction wrought there by the Spanish. Similarly, when Cortés at the end of his second letter compared Tenochtitlán to Jerusalem, he was not thinking of the holy city desired by the crusaders. For Cortés, the two cities were comparable only in their justifiable destruction by divine retribution.[22] The New World, for Columbus's successors, was truly new: it was neither Jerusalem nor a paradise lost. Its conquest could not, therefore, be considered a recovery of lost spiritual grounds, and it could not be a crusade in the traditional sense. And although the crusading theme did much to shape their attitudes toward America and legitimize its conquest, it would clearly not suffice: other conceptual resources and role models were needed. They were found in the literature of romance and travel.

Crusaders, as it turned out, were not the only traveling warriors of the early modern European imagination. The knights-errant of medieval romance were enjoying a surge in popularity at precisely the time their real-life counterparts were attempting the conquest of a new world.[23] The heroic deeds of the Knights of the Round Table, as well those of their now forgotten counterparts like Amadis of Gaul and Palmerin of England, were being recounted in an ever-growing body of literature. Both old and new romances were now accessible to an ever-growing readership, thanks to the invention of the printing press. The analogy between the deeds of the fictional heroic knights and their real-life counterparts in the Americas was obvious to the reading public of the time. As Jennifer Goodman points out, the same printer who published Cortés's *Cartas de Relación* (*Letters from Mexico*), recounting his conquest of Mexico, simultaneously published Jean Bagnyon's *Historia del Emperador Carlo Magno*, which recounted the deeds of Roland and his companions in Moorish Spain.[24] Both accounts could be found side by side on booksellers' shelves.

The fictional knights of the romances appear at first sight to share much of the ethos and attitudes of the historical conquerors of the New World. Many romantic heroes were presented as soldiers of God, an ideal that accorded well with the crusading self-image of the conquistadors. Some, like John Mandeville, traveled to Jerusalem, and Huon of Bordeaux even liberated the holy sites. Those who did not actually make the pilgrimage were sometimes assigned other religious tasks, such as King Arthur's knights' search for the Holy Grail.[25] Some, like Sir Galahad, were exceptionally saintly, and in general all were willing to lay down their lives for the defense of the faith. The romances, it is clear, borrowed unhesitatingly from crusading literature and ethos.

But although Huon of Bordeaux and John Mandeville traveled to Jerusalem, they were not crusaders. The Holy City was not the end of all ends for them, but rather a way station on the road to ever more wondrous adventures. They moved, in Greenblatt's terms, from "the center of the earth" to "the rim of the world."[26] Huon, for example, after traveling to Jerusalem, moved on to Babylon with his betrothed, the Saracen princess, by traveling through a magical wood and aboard a sea monster. He ultimately returned to Bordeaux only to be imprisoned by his wicked brother Gerard.[27] Mandeville, after spending the first half of his book traveling toward Jerusalem, then moved on to the exotic lands of the East, where he recounts seeing many wonders and great riches. He too ultimately returned home to England.[28] For Columbus, the far reaches of the earth were merely a stop on the way to Jerusalem. In a precise inversion of this view, for the romantic knights-errant, Jerusalem itself was a stop on the way to "the rim of the world."

The fundamental difference between crusaders and knights-errant was in the purpose of their respective journeys. The crusader was single-mindedly goal oriented: his mission was to reach his Promised Land and possess it. All else was merely a means, subject to this end. No matter how far from his goal his journey will take him, it must be considered part of the road that leads indelibly to the Holy City. For the knight-errant, in contrast, the journey is the aim in itself. The oceans he crosses, the forests he travels through, the mountains he climbs, and the valleys he traverses do not really lead anywhere. They are simply part of his never-ending quest, which leads nowhere except to more journeying. Illusory targets may appear along the way: Jerusalem may need to be saved, or the Holy Grail recovered, or great treasures unveiled.[29] But Jerusalem is soon abandoned, and the Holy Grail remains perpetually beyond reach, serving as a spur for further wanderings. The knight-errant forever continues on his journey, dueling with other knights, saving maidens, and recovering treasures as they come his way.[30]

The influence of the chivalric romances is everywhere in evidence in the exploration literature of the New World, but a few examples here are helpful.[31] Explorers of the Americas, from Cortés in the 1520s to Walter Raleigh in the 1590s, persistently sought to locate the "island of the Amazons," which was a common theme in the romance literature. In a similar vein, Bernal Díaz, chronicler of the conquest of Mexico, compared the great city of Tenochtitlán to "the enchanted things related in the book of Amadis."[32] Elsewhere he expressed his hope that "God give us the same good fortune in fighting . . . as he gave the Paladin Roland."[33] Cortés was also a great admirer of that paragon of chivalry and in his letters sought to fashion himself as the "Roland of the Americas." In his own eyes, he was fighting to enlarge the empire of Charles V just as Roland had done for Charlemagne.[34]

The relationship is perhaps most pointed in the case of Gonzalo Fernandez de Oviedo, who was Las Casas's enemy and leading apologist for the deeds of the conquistadors in the Indies. Oviedo was the author of not only the widely acclaimed *Historia general y natural de las Indias* but also a lesser-known work recounting the deeds of the valiant (and fictional) knight-errant Claribalte.[35] Oviedo, in other words, was at once an eyewitness to the exploration and conquest of a new world and the author of a romance of chivalric heroic wanderings. It is less than surprising if his idealized view of the conquistadors in the Indies came to be colored by the heroic imagery and conventions of the romance genre.

It is not, however, only direct literary references that point to the close relationship of exploration literature to the chivalric romances. The very landscape of the discovery and the explorers' place within it seem to be closely patterned on the imagery of romance. Jennifer Goodman points out how the Mexico of Cortés's *Letters from Mexico* resonate with the landscape of Moorish Spain, depicted in the saga of Roland. The difficult passages through the mountains, the bridges bravely defended, and the wondrous alien city in the interior of Mexico are all reminiscent of Roland's adventure.[36]

Even in the absence of such a direct correlation between specific texts, the imprint of the terrain of romantic adventure on the landscape of early America is impossible to avoid. The long voyage across the seas in search of fame and glory is as common a theme in chivalric tales as it is, inevitably, in New World exploration narratives. Once on land, the explorers encountered great mountains, as did Cortés, or deep forests, as did those who sought the lost kingdom of El Dorado in the South American jungle later in the century.[37] They traveled through mountainous passes and along great rivers and, with divine guidance, ultimately overcame the obstacles in their path. In this they were inspired by the feats of Roland and his companions in the rugged Pyrenees, or the Knights of the Round Table in their sylvan wanderings. Both New World explorer and medieval knight-errant traveled through uncharted lands and overcame forbidding natural obstacles and dangerous human foes in search of an elusive prize. It is hardly surprising that the explorers, faced with the unprecedented novelty of the New World, sought to fashion themselves on the familiar and well-established images of chivalric literature.

How then is one to characterize the exploits of the early explorers of the Americas? Were they crusaders? Many undoubtedly viewed themselves in such terms, and to an extent they indeed followed in the crusader's footsteps. Their single-minded determination to occupy and possess their promised lands in the name of God and to spread His word among the heathen are hallmarks of crusading. And yet it was an odd crusade indeed. Rather than aiming to retrace one's steps to the pure origin at the center of the world, the explorers took their mission to the ends of the earth, to uncharted and unheard-of lands.

Were they, conversely, real-life knights-errant in search of their Holy Grail? Again, there is much evidence to support such a view: both consciously and unconsciously, the explorers modeled themselves and their deeds on the feats of the romantic heroes. Like them they traveled to distant unknown lands and overcame great hazards in their pursuit of elusive goals. There are, however, important differences here as well. The romantic hero was a perpetual traveler whose quest never ends, because his goal shifts or moves further away the closer he draws to it. He travels through unknown lands, but he does not "discover" them. After his passage the lands remain as unknown and mysterious as they were before he ever set foot on them. His voyage is ultimately a voyage of self-discovery, a chance for his latent nature as a courageous and chivalrous knight to express itself. The wondrous landscapes he travels through are simply an opportunity for the knight to exhibit and perform his chivalrous nature.

Not so with the New World explorer, who is driven to discover, map, and possess the newfound territories. His efforts may reveal the truth about his hidden nature, but they are also single-minded attempts to claim the lands for himself and his sovereign. Huon of Bordeaux never claimed "Baghdad" for himself or for France, and Mandeville never attempted to plant the seeds of an English empire in the many lands he visited. But Columbus claimed the Indies for Spain, Cortés did the same in Mexico, and Walter Raleigh attempted (and failed) to turn Guyana and its reputed riches into an English colony. These explorers may have shared the romantic literature's vision of hazardous travel through distant enchanted lands, but when it came to the ultimate meaning and purpose of their exploration they traveled a very different path.

The American explorers were neither crusaders nor knights-errant, but rather drew heavily on both traditions. They co-opted the romantic landscape of the chivalric tales and saw themselves as traveling through a faraway enchanted land of towering mountains and impenetrable forests in a quest for honor and riches. Like their fictional heroes, they believed they would ultimately find their way through all obstacles and emerge victorious from whatever challenges come their way. At the same time they drew on the crusaders' sense of divine mission and their single-minded drive to conquer, possess, and master the lands they targeted. The New World explorers redirected the zeal of soldiers of the cross from the center of the world to the far reaches of the earth, the New World.

The result was a narrative of exploration that drew on older models but combined them in a way that was unmistakably new. In this tale, the new lands took on many of the characteristics of the landscape of romance, while the explorer himself acquired many of the traits of the crusader. In fine romantic style, America was most often presented as a dangerous and forbidding land of mountains and forests, but hiding within its fold a great prize for heroes to pursue. That prize might be a great city, like Tenochtitlán, an enormous golden mine, like El Dorado, or simply wondrously bounteous land, such as Virginia.[38] Like the knight-errant of romance, the explorer who arrived in this new land would seek to find his way to the bounties hidden within it. But where the knight-errant would settle for a mere taste of these riches, leaving them much as he had found them, the crusading explorer would insist on demolishing all obstacles in his way and ultimately occupying and possessing the land of his desire.

The following chapter shows how these themes came together to form a standard narrative of exploration and discovery, which would be repeated in different forms throughout the age of exploration. It posited a wondrous land of riches in the interior, surrounded by hazardous terrain of forests, mountains, and, occasionally, icebergs. The enterprising explorer who arrived on its shore would find hidden passages and break through all obstacles in his way to the fabled land of the interior. For this he was rewarded with fabulous riches and the possession of a wondrous land. It was a tale of both romance and crusade, but in itself was neither the one nor the other. It was a new tale of exploration and discovery, different from the stories and myths that came before it. It would serve as an inspiration to explorers throughout the age of discovery, ultimately having profound consequences for the emerging geography of the New World.

CHAPTER 2

Breaking Alien Coastlines

A GEOGRAPHY FOR A NEW WORLD

At the end of his detailed account of the voyages of Martin Frobisher in search of the Northwest Passage, George Best summed up Frobisher's conclusions regarding the topography of North America: "It is now found," he wrote,

> that Queene Elizabeths Cape being situate in latitude at 61 degrees and a half, which before was supposed to be part of the firme land of America, and also all the rest of the South side of Frobishers straits, are all severall islands and broken land, and likewise so will all the North side of the said straites fall out to be as I thinke. . . . These broken lands and Islands being very many in number, do seem to make there an Archipelagus.[1]

Frobisher and Best were undoubtedly wrong. The area explored by Frobisher was the southern part of the east coast of Baffin Island and Hudson Strait, which separate it from the North American continent to the south. Best's "archipelagus" appears in modern maps as nothing more than shallow promontories and bays along the coast of Baffin Island. Nothing remains of the islands that lined "the way to Cathaia" in James Beare's map of the expedition.[2]

But although Frobisher and Best were mistaken, they were not alone in their views. John Davis led three expeditions to the same region in the 1580s, several years after Frobisher, and his account of his voyages was published in Hakluyt's *Principal Navigations*. While sailing north along the coasts of what is now known as Davis Strait, between Greenland and Baffin Island, Davis reported that he "searched still for probabilities of the passage, and first found, that this place was all islands, with great sounds passing betweene them."[3] The return voyage south brought similar results: "As we coasted the south shore we saw many faire sounds, whereby we were perswaded that it was no firme land but islands."[4] On a different occasion the explorers' attempted to "sail into a mighty great river directly into the body of the land," only to find it "to be no firme land, but huge waste, and desert Isles with mighty sounds and inlets passing betweene sea and sea."[5] When Davis proceeded to climb "a very high hill," he "perceived that this land was islands."[6] At the end of his account, Davis reiterated his seemingly inevitable conclusion: "for no doubt" he wrote, "the North partes of America are all islands by ought that I could perceive therein."[7]

Frobisher and Davis were among the most famous English sailors of their age and prided themselves on their navigational skills and accuracy of observation. They repeatedly emphasized the veracity of their descriptions and took great pains to transport to England material proofs of their discoveries.[8] Indeed, even their description of the Canadian arctic was rooted in geographical reality, since these coasts are marked by numerous small islets. The surprising thing about Frobisher's and Davis's reports is not that they found many islands, but rather that they completely missed the North American landmass. The massive bulk of the continent of North America was carved up in their maps and narratives into a complex web of islands, narrows, and straits (see Figure 1).

Other English explorers of this time came to similar conclusions. Humphrey Gilbert, for example, published a map in 1576, which, although less detailed, strongly resembled Frobisher's later chart (see Figure 2). For Gilbert, as for Frobisher and Davis, the northeastern coast of America turned out to be composed of islands and straits. Explorers of more southerly regions displayed similar, if less extreme tendencies. Although they did not reduce the entire North American continent to small islands like their northerly contemporaries, they did consistently break down the surveyed coastal regions into islands and passages. Philip Amadas and Arthur Barlowe did so with regard to the coast of Virginia when they traveled there in 1584. Their account was confirmed by Thomas Hariot several years later and restated as late as 1625 in a treatise by Henry Briggs. Walter Raleigh and Lawrence Keymis explored the coast of Guyana in search of El Dorado in the 1590s, and they too found that the coastline was dissected by innumerable rivers, inlets, and small islands.[9] Occasionally, large new islands were created in the process; such was the case with Norumbega, which occupied the approximate area later known as New England in maps of the period.[10]

A peculiar tendency thus emerges in the exploration narratives of the Elizabethan age: time after time, English explorers and mapmakers dissolved the land they were surveying into islands, straits, and rivers. Baffin Island, Virginia, and Guyana were all subjected to this practice. The carving up of the newly discovered lands thus became a hallmark of the maps and written

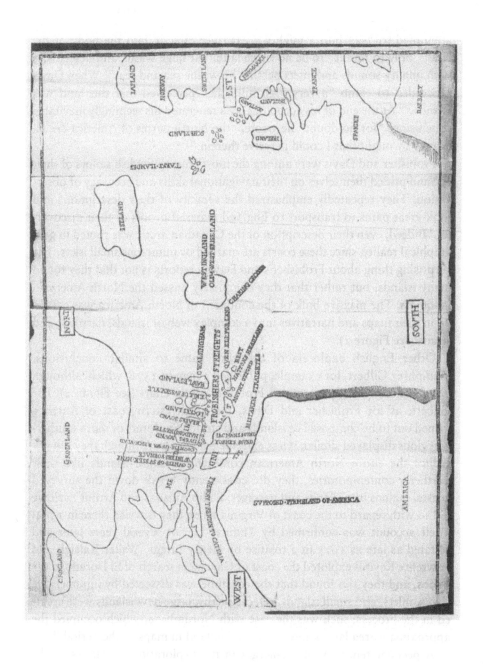

accounts of the English western enterprise. The practice is indeed a puzzling one. Why, it may be asked, did some of the best navigators of the age come to (mis)represent the lands they surveyed in this particular manner? The shortcomings and limitations of surveying techniques in the sixteenth century clearly played a role in the explorers' errors, but this is not a sufficient explanation. Technological deficiencies may account for inaccuracies and straightforward mistakes in the explorers' reports. They cannot, however, explain the consistency with which travelers in different settings and with diverging purposes described the new lands in such similar terms.

The answer, I will argue, lies in the simple tale of exploration and discovery that pervaded Elizabethan voyagers' views and accounts of the lands through which they traveled. It should, of course, be familiar by now, for it was the very tale that combined elements drawn from the medieval crusading tradition with others drawn from the literature of romantic chivalry. It posited, on the one hand, a heroic traveler who was in many ways a crusader, out to conquer and possess his own "Promised Land." The land itself, on the other hand, seemed to be drawn from the exotic and magical landscape of romance rather than the sacred geography of a crusade. It was wondrous and magical, concealing many untold secrets within its fold. Most importantly, it lay at the ends of the earth, completely unknown and uncharted, untouched by a Christian or European past. The combination of a crusading explorer with a romantic landscape brought about a unique story of exploration and discovery, which pervaded the travel narratives of the time.

This narrative informed the explorers, before they ever set sail, about what they might expect to find in the foreign land and why it was worthwhile to pursue it. It outlined the type of difficulties that would be encountered in the process and the ways in which they should be overcome. When Frobisher, Davis, and Raleigh actually encountered the alien lands, they already knew what to look for, where to find it, and how to get there. Their maps and accounts were not simply reflections of their experiences, they were also reports on their progress along a preknown route, on their way to goals set out in advance.[11] Their accounts were not just tales of actual exploration of unknown geographical regions but also of travel in the well-known but imaginary geography outlined in the Elizabethan story of discovery.

Figure 1 (opposite).   James Beare(?), Map of Frobisher's Voyages. Printed in George Best, *A True Discourse of the late voyages of discouerie, for the finding of a passage to Cathaya, by the Northweast, under the conduct of Martin Frobisher Generall* (London, 1578) (Courtesy of the Huntington Library).

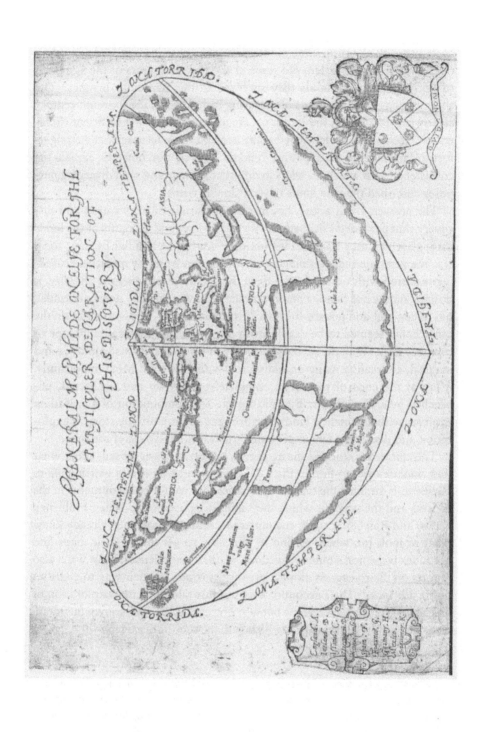

The main outlines of this story of discovery are simple: deep within the heart of the country, the tale goes, is a hidden land, rich in gold and many treasures. It lies far in the interior, protected by mountain ranges and other natural barriers. It is often depicted on the shores of a lake, or, perhaps, on an island. It is remote and inaccessible and appears completely impenetrable to the traveler who arrives on its hostile shores. The enterprising explorer, however, must not despair: the apparently impenetrable coastline is, in fact, lined with straits and rivers, some of which break through all barriers and provide passage to the golden interior. If he is brave and resourceful, the explorer will enter the passageways, overcome all obstacles, and find his way through to the hidden land. Once there, he will be rewarded with immense riches.

The English explorers were so familiar with this tale that it structured their very conception of their enterprise and pervaded their reports. As a result, when they were faced with the completely unknown coasts of America, they regarded themselves as surprisingly well informed. Following the story, they fully expected to find stores of gold and riches in the interior. If these proved difficult to locate, then they could attribute this to the fact that they were hidden and protected by the natural terrain. This, of course, was also quite consistent with their expectations. Most importantly, the explorer's own role was defined in the story: he was to stubbornly attempt to break through all obstacles, doggedly and persistently locate the passages that may be found along the coastline, and penetrate the interior. Every coastal inlet sighted was seen as a potential strait leading inward, and every bay became a possible passage to the golden interior. If one considers this guiding narrative, it seems quite understandable that the lands surveyed were transformed in the explorers' accounts into a maze of straits and passageways, separated by disconnected islands. The explorers' accounts attempted to bridge the gap between the familiar master story of exploration and the stark, unaccommodating realities of the foreign coasts. In doing so, they ended up carving the actual landscape with imaginary passages.[12]

In drawing their maps of newly discovered lands, Elizabethan explorers achieved a dual purpose. First of all, needless to say, the maps were meant to be accurate records of the voyagers' findings, which would facilitate further exploration of the regions. As such, they were largely what they claimed

Figure 2 (opposite).   Sir Humphrey Gilbert's world map. Printed in Humphrey Gilbert, *A Discourse of a Discoverie for a New Pasage to Cataia* (London, 1576) (Courtesy of the Huntington Library).

to be—direct depictions of foreign lands using conventional signs, accurate to the degree allowable by the technology of the time. At the same time, however, the maps also served as mythical and ideological constructs, reifying the tale of exploration and discovery and writing it onto the alien landscape. When presented in cartographic form, the credibility of the tale of exploration was greatly enhanced. As a map, it was no longer a hope or aspiration, but simply part of the natural landscape of the new lands. A map, after all, presents itself as nothing more than a direct description of natural terrain.

Achieving both these goals simultaneously was, of course, a difficult and paradoxical proposition. How can a single chart serve as both a more or less accurate depiction of a foreign land and as a representation of a mythical tale? Following Dennis Wood, I suggest that the maps produced by the explorers should be viewed as "myths," in the sense used by the French theorist Roland Barthes. A myth, according to Barthes, is a two-tiered system, composed of a basic level, which he calls "language," and a superimposed level of "metalanguage" that makes use of "language" to compose the "myth." He gives the example of a magazine cover from the 1950s sporting the photograph of a black African in French military uniform, saluting the tricolor. On the level of "language," this is simply a depiction of a particular man in a specific situation; as "myth," however, it is an ideological representation of the French Empire. Anyone viewing this fine soldier, it suggests, would easily see that Imperial France treats and serves all its peoples alike, regardless of origin or race. The advantage of this two-tiered arrangement, according to Barthes, is that in the form of a myth, ideological claims can masquerade as nothing more than a simple straightforward reality. What seems like a dubious claim when presented in itself—that the French Empire is an egalitarian institution that commands the loyalty of all its subjects—becomes a simple statement of fact when presented as myth. How could anyone doubt it? The zeal with which this African soldier salutes the flag of his supposed oppressors speaks for itself![13]

The case of the exploration maps is structurally similar to that of the magazine cover. On the level of "language," the maps are nothing more than direct representations of the lands they depict. As myths, however, they signify the ideological tale of exploration and discovery, the promise of great riches and the heroic quest in the face daunting difficulties. The advantage of this mythical structure is clear: if the ideological claims of the map are challenged, it will always fall back on its function as a transparent depiction of the contours of the land. For what could be open to dispute in an accurate diagram of natural features? By functioning simultaneously on the level of language and of myth, the maps naturalize the ideological claims of the rhetoric of exploration and present them as parts of the geographical landscape.[14]

In demonstrating the narrative basis of early modern English imperialist geography, I will be concentrating on four major ventures of the Elizabethan and early Stuart periods. The first is the search for the Northwest Passage to Cathay, undertaken by Martin Frobisher in the 1570s and followed up by John Davis in the 1580s. Second is Walter Raleigh's attempt to found a colony in Virginia between 1584 and 1587. Third is his search for the golden kingdom of El Dorado in Guyana, which began in 1595 and was to cost Raleigh his life twenty-three years later. Last of all is the renewed effort to locate the Northwest Passage under the early Stuarts. In this context I will discuss Henry Briggs's treatise promoting the possibility of a river passage through Virginia. Each of these cases reveals the presence of the underlying standard narrative of exploration, shaping the views of the explorers and their accounts of the newfound lands. Specifically, I will point out the ways in which this simple story helped shape a unique geography of exploration, dissecting the coastlines of America with rivers, straits, and passageways.

THE UNKNOWN MARK: FROBISHER AND DAVIS IN THE CANADIAN ARCTIC

The existence of a northwest passage to Cathay along the north coast of North America was a popular notion among West European cosmographers in the sixteenth century. The Portuguese domination of the eastern route around the Cape of Good Hope and the Spanish claims over the Straits of Magellan left other European nations searching for ways to circumvent the Iberian monopoly. Several expeditions set out from England beginning in the 1550s in search of a northeast passage around the north coast of Asia. While failing in their original intent, the voyages did, nonetheless, result in the opening of trade relations with the Russian Empire. Later in the century the focus shifted to the northwest. Mercator's 1569 world map, which showed a clear passage separating northern "New India" from a hypothetical Arctic continent, was well known in England. So was Ortelius's *Theatrum* of 1570, which included a map showing a long strait, stretching from "Estotiland" in the east to the "Straits of Anian" in the West.[15] It is hardly surprising that, finding the southern routes to the East closed, English imperialists should focus on finding a northern route around them.

Martin Frobisher's three voyages between 1576 and 1580 were the first systematic attempts by the English to search out the reputed passage. His first expedition was the smallest of the three and included only two small ships of 30 tons—the *Gabriel* and the *Michael*—and a pinnace.[16] The small fleet set sail on June 7, 1576, stopping first at the southern tip of Greenland, which they took to be the island of "Friezland." Continuing westward, they arrived at the southern part of Baffin Island late in July. Turning north along the coast, they entered an inlet two weeks later, which they took to be the eastern entrance of the Northwest Passage, and named it accordingly Frobisher's strait (Frobisher Bay to future generations). The next several weeks were spent exploring the passage and establishing contact with the local Eskimos with rather unhappy results: five of Frobisher's men were taken prisoner by the locals, and one Innuit man was abducted by the English. Having failed to secure a prisoner exchange, Frobisher turned back east at the end of August and arrived in England early in October.

Frobisher's assertion that he had discovered the entrance to the Northwest Passage helped him raise the necessary funds for a follow-up expedition the next year. Even more important was the promise of gold: Frobisher brought back from America a piece of black rock, which he hoped would prove auriferous. Although several assayers found nothing of value in the rock, one was eventually found who declared it to be laded with gold. This find created such an impression that even the usually skeptical queen was persuaded to invest in the venture and place one of her own ships under Frobisher's command in the second voyage. A lingering doubt was, nonetheless, evident in the name Elizabeth gave the newly discovered land—"Meta Incognita"—the unknown mark.

The second expedition, which included the queen's ship *Aid* in addition to the smaller *Gabriel* and *Michael*, set out from England in May 1577. The original goal of the enterprise had now been pushed to the background: instead of attempting the discovery of a passage to Cathay, the main purpose of the second voyage was the mining of gold in Meta Incognita. As before, the expedition first sailed around the southern tip of Greenland, and then proceeded west to Frobisher Bay. During the following three weeks Frobisher and his men excavated massive amounts of the red and black rock and loaded them onto the ships in the hope that it contained gold. Several skirmishes with the locals resulted in the capture and imprisonment of an Innuit man and woman, who were brought back to England but died shortly after their arrival. On August 23 the three ships started back on their way home, arriving in Wales on September 23 with nearly 200 tons of ore.

Following the return of the second expedition, the queen appointed special commissioners to inspect the ore and evaluate its gold content. Although the results were inconclusive, a third voyage was organized on the assumption that gold had indeed been found. The fleet, this time, was far larger than its predecessors, composed of no less than fifteen ships. Its goals were considerably more ambitious as well: in addition to the extensive mining operations, the expedition was also instructed to plant a small settlement to guard the mines, explore the land, and continue to search for the elusive Northwest Passage. With these lofty goals in mind, the ships set out from Harwich on May 31, 1578.

After the customary stopover in southern Greenland, Frobisher's fleet came in sight of Meta Incognita on July 2. The approaches to Frobisher Bay were, however, blocked with ice, and the storm that broke out shortly afterward greatly compounded the danger. For several days the ships were swept south until, on July 7, they entered a great passage that opened westward, later to be known as Hudson Strait. Convinced that he was at the entrance of his long-sought passage, Frobisher pushed westward for nine days, declaring to his increasingly skeptical company that they were in fact in Frobisher's strait. Eventually the fleet turned back, and breaking through the ice, entered Frobisher Bay in late July.

At this point the ambitious goals of the expedition were cut back drastically. Two prefabricated walls of the intended permanent fort were lost at sea, and the plan to leave a small colony on the land was quickly abandoned. No further exploration of the Northwest Passage was attempted either. Only the mining operations proceeded as planned, and the ships were loaded with red and black rock. Before leaving, the explorers built a small house in which knives, pictures of men and women on horseback, and others of our "country toyes" were left for the benefit of "those brutish and uncivill people."[17] The ships then sailed home on the last day of August, arriving in England late in the year. In the following months the assayers could not produce the promised gold, and support for the enterprise eroded quickly. Frobisher's third voyage to the North American arctic was also his last.

The search for the Northwest Passage was next taken on by John Davis "The Navigator." Between 1585 and 1587 Davis led his two small barks, the *Sunshine* of 50 tons and the *Moonshine* of 35 tons, on three extensive surveys of the Northwest. Like Frobisher, he first sailed to "Friezland," which, unlike his predecessor, he correctly identified as the southern tip of Greenland. Sailing north along the western coast of the island, he entered the narrows separating Baffin Island from Greenland and named them after himself— Davis Strait. He reached as far north as 73 degrees, where he ran into pack ice and was forced to turn west to Baffin Island. Beyond the ice, he claimed, he could see an open expanse of sea; the Northwest Passage, it seemed, was within easy reach. On his way back south Davis noted the entrance to

Hudson Strait but did not investigate. The sought-after passage, he believed, would be found to the north, in the open sea beyond Davis Strait.[18]

In the months following the voyages, the participants in Frobisher's and Davis's ventures produced an impressive array of firsthand accounts of the expeditions. Every literate seaman, it seems, was anxious to disseminate his own experiences to an eagerly awaiting public. Although differing in personal style and perspective, all accounts were united by a central theme. Underlying the detailed descriptions of their respective adventures was the consistent and familiar framework of the standard narrative of exploration. This simple story structured the participants' accounts of their experiences and shaped their perception of the newfound land.

For the sake of convenience, I have divided this tale into its three central components: first is the "golden land" theme. This part sets the expeditions in motion by positing great riches that will be gained by penetrating the undiscovered country. The legend of El Dorado—a nonexistent country named after and defined by its gold—is probably the purest form of this theme. The second theme deals with the difficulties encountered in approaching the golden interior. This part is a somewhat inconsistent element in the narratives, since although the explorers usually emphasize their own difficulties and sufferings in pursuing their goal, they may choose, on occasion, to downplay the obstacles in order to attract money and volunteers to their ventures. I have chosen, nonetheless, to include it in my analysis since it is a major element in the tales of the participants themselves, as against the pamphlets of the promoters who stayed behind.

Even more importantly, the description of the difficulties sets the stage for the third and crucial theme—the passages to the interior. The obstacles and barriers along the way to the "golden interior" emphasize the need for clear unobstructed passages, which will cut through them. It is the plethora of such passages, some real, more often imagined, which results in the carving up of the land with rivers and straits.

Let us begin with the first theme of the exploration narrative—the hidden gold that starts the expedition on its way. The original "golden land" of the search for the Northwest Passage was, of course, Cathay—the general term used by the explorers to refer to China and its neighboring lands. "You might justly have charged me with an unsettled head if I had at any time taken in hand to discover Utopia," wrote Humphrey Gilbert in 1566. "But Cataia is none such, it is a countrey, well knowen to be described and set foorth by all modern geographers." It is a country, he adds, "whose abundance of riches and treasure, no man of learning, and judgment doubteth, for that the countreys themselves, and their commodities, are apparently known by sundrie men of experience."[19] Gilbert was, of course, repeating a long-standing Western tradition, easily traceable to classical times and familiar to Elizabethans through the writings of Marco Polo and John Mandeville.[20] By discovering the Northwest Passage to Cathay, Gilbert claimed, "we may sail to divers marveilous riche Countries both Civill and others, out of both their jurisdictions, trades and traffiks, where there is to be found great aboundance of gold, silver, precious stones, Cloth of golde, silkes, all manner of Spices, Grocery wares, and other kindes of Merchandize, of an inestimable price." "Ye wealth of all the East partes of the worlde," he reiterated, "is infinite."[21]

Such was the vision behind the northwest voyages of Frobisher and Davis: Cathay was a legendary land whose very name became synonymous with great riches. Nothing was known about this mysterious country, except that its discovery would be rewarded with immense wealth. Michael Lok, Frobisher's main financier, thought he need say nothing "touching the naturall riches and infinit t[reasure?] and the gret trafik of rich merchandise that is in those countries of Kathay, China, India" in his account of the first voyage. These facts, Lok considered, are universally known.[22] George Best, in the "Epistle Dedicatory" to his own narrative of Frobisher's voyages, agreed: "The passage to Cataya," he wrote, "promisseth us more riches by a nearer way than either *Spaine* or *Portugale* possesseth."[23]

The unique feature of Frobisher's voyages is that this vision of Cathay became displaced during the expedition itself. When the passage to the East proved to be more difficult than was at first hoped, the Orient and its riches were projected onto the barren land that happened to be at hand. It was indeed a strange choice: the desolate frozen landscape of the Canadian arctic and its Innuit inhabitants could hardly be more different from the great civilizations of the East described by Mandeville and Marco Polo. This notwithstanding, Meta Incognita was speedily transformed in Frobisher's project, from a troublesome obstacle on the way to Cathay into a legitimate target in its own right. This was accomplished by the convenient discovery in Baffin Island of the commodity that epitomized the goals of the discovery voyages: gold.[24]

George Best gave an obviously apocryphal account on how gold was discovered in Meta Incognita. When Frobisher returned from his first voyage, Best wrote, his friends asked him what he had brought with him from "that countrey." Since he had nothing left at hand, Frobisher gave them a piece of black stone. As it happened, the wife of one of these friends threw a piece of the stone into a fire. When it was withdrawn, it "glistered with a bright marquesset of golde."[25] Michael Lok, the chief financier of the voyages, related a somewhat more likely version of events, although his story too has strong mythical overtones. Frobisher, according to Lok, promised him before first setting sail to give him the first thing that he found in the new land. Upon his return, Frobisher invited Lok to his ship and, in the presence of witnesses, handed him the black stone. Lok wasted no time and immediately sent the rock to be assayed for gold content. Eventually an assayer was found who confirmed what Lok had suspected all along: the rock was, indeed, auriferous.[26]

The purpose of Best's and Lok's apocryphal stories is clearly to stress that the discovery of gold was accidental and uncontrived. In Best's version, an unlikely chain of coincidences is created in order to distance the discovery from any purposeful machinations, which could be attributed to Frobisher. The rock was not brought over for assaying, but was merely a piece of debris. It was all Frobisher had at hand when his friends came to ask for items from the new land. As a result it was not in Frobisher's possession when gold was discovered, but in that of his associates, and even then the discovery was made by an unsuspecting woman who unwittingly dropped a piece of the rock in the fire. Lok's story is simpler and somewhat more plausible. The insistence on coincidence is, nonetheless, apparent in Lok's assertion that Frobisher did not actively seek the rock but that it was simply the first item he encountered in Meta Incognita. The randomness of the discovery of gold points to its fantastic abundance in the unexplored land: even mere rocks and debris on Meta Incognita, the first things Frobisher happened to come across, prove to be golden. At the same time, the elaborate chain of coincidences leading to the discovery of the gold is designed to insulate the finding from the desires of the explorers. The gold, the tales imply, revealed itself independently of any human agency. Its existence should not, therefore, be doubted.

The defensive tone of the stories is due, no doubt, to the fact that a great deal of human agency was indeed required in order to discover gold in the rocks. Lok's letters reveal his untiring determination to send the rock to one assayer after another until one was eventually found who would confirm the presence of gold.[27] As the apocryphal tales suggest, the claim was still viewed as rather shaky when Lok and Best were writing. The suspicion seems to have persisted among potential backers that the gold was a reflection of the adventurers' desires rather than an overseas reality. The skeptical public, which was no doubt familiar with wondrous tales of exploration, suspected that it was being subjected to a reiteration of a familiar story of discovery, rather than a true report.

Even for the explorers themselves the presence of gold in the barren rocks of Meta Incognita was no settled matter. Dionise Settle's account of the second voyage sounds a decided note of caution regarding the "supposed" gold ore; Upon entering Frobisher's strait, Settle reports, Frobisher set out with a group of men to search the country. "The day following, beinge the 19 of

Julie our captain returned to the ship with report of supposed riches which shewed itself in the bowels of those barren mountaines, wherewith we were all satisfied."[28] Several days later the fleet had sailed further up the bay, "Where our Generall [that is, Frobisher] had found out good harbour for the ship and barks to anker in, and also such store of supposed gold ore as he thought himself satisfied withall." Thus far, it seems, Settle attributes the search for gold almost solely to Frobisher personally. The author himself, it seems, remained somewhat cautious regarding the "supposed" riches.

Later in the account Settle grows more optimistic: the Eskimos, he relates, "make signes of certaine people that weare bright plates of gold in their foreheads, and other places of their bodies."[29] The riches are now not merely Frobisher's personal obsession, but are attested to independently by the locals. In his general summation of the findings of the expedition, Settle takes the position of guarded optimism. Although cold and frost is, in his opinion, detrimental to the formation of gold, beneath this external crust "the earth within is kept the warmer, and springs have their recourse, which is the only nutriment of golde and Minerals within the same." Gold and minerals, in Settle's view, lie hidden beneath the arctic terrain, nourished by air and water, awaiting their discoverer.

Nevertheless, he ends on a note of caution: "There is much to be sayd of the commodities of these Countreys, which are couched within the bowels of the earth, which I let passe till more perfect triall be made thereof."[30] George Best appears on the whole to be more optimistic than Settle with regard to the prospects of discovering treasure. Even before arriving in Frobisher Bay on the second voyage, Best was already watching out for potential riches. On the stopover in "Friezland" he observed "a kinde of Corall almost white, and small stones as bright as Christall." He then quickly concludes, "And it is not to be doubted but that this land may be found very rich and beneficial if it be thoroughly discovered."[31] Best clearly did not require much evidence in order to find hidden riches. The mere existence of an unknown land was for him an indication of great wealth in the interior.

Best continues in the same optimistic vein when he arrives in Meta Incognita itself: "Upon a small Iland within this sounde called Smithes Iland . . . was found a Mine of silver . . . here our goldfiners made say of such Ore as they found upon the Northerland, and found foure sortes thereof to hold gold in good quantitie."[32] Best is evidently not a doubter. The same rocks that Settle thought required further testing, seemed much more promising to Best. The discovery of silver is reported as simple fact, and the presence of gold is indicated by the expert assaying of the ore. The "golde ore," Best adds in his account of the third voyage, "had appearance & made shew of great riches and profit." It is no longer a problematic unresolved issue, but rather a discovery already made. The gold is placed securely deep within the land of Meta Incognita, and the mental transfer of riches from Cathay to Baffin Island is thus complete.

The search for gold and riches was, of course, implicit in the very nature of the northwest enterprise. It was the obvious motive that drove the entire project and is therefore stated only in the broadest programmatic pamphlets, such as Gilbert's. There was no need to repeatedly remind the readers of the travel account that the attainment of great wealth was a primary goal of the expeditions. The reason why these issues were broached in Settle's and Best's accounts was that the promised gold was shifted from Cathay to Meta Incognita. Settle and Best were obliged to justify this move and to establish the new land as a legitimate target for the voyages, hence their discussion of "the golden interior."

The case is quite different with regard to the second motif of the exploration narrative, which is the very heart of the travel tale: the obstacles surrounding the golden interior. The great difficulties of the road to the promised land, the hardships suffered, and the losses incurred make up much of the substance of this kind of story. Such indeed was the case with the chroniclers of Frobisher's voyages. If the theme of "gold in the interior" is something of an afterthought, brought about by the change of plans, the discussion of the obstacles barring the way to those riches is central to the accounts.

The main outlines of the "obstacles" theme are clearly in evidence in Settle's description of the coast of "Friezland" in the second voyage:

> All along the coast yce lieth as a continuall bulwarke, & so defendeth the
> Countrey, that those that would land there, incur great danger. Our Generall
> 3 dayes together attempted with the ship boate to have gone on shoare, which
> for that without great danger he could not accomplish, he deferred it untill a
> more convenient time. All along the coast lie very high mountaines covered with
> snow, except in such places, where through the steepenes of the mountaines of
> force it must needs fall. Foure dayes costing along this land, we founde no sign
> of habitation. Little birds, whiche we judged to have lost the shore, by reason
> of thicke fogges which that Countrey is much subject unto, came flying into our
> ships, which causeth us to suppose, that the countrey is both more tollerable, and
> also habitable within, than the outward shore maketh shew of signification.[33]

The coast of Friezland, according to Settle, is naturally fortified. A "continuall bulwarke" of ice protects the approaches and "defendeth" the country, and "thicke fogges" further compound the difficulties of the approach. "High mountaines" along the coast provide a second line of defense against anyone who would make it that far. This natural fort is, appropriately enough, assaulted by a "Generall," who fails to break through its natural fortifications. The coastline is continuous, impenetrable, and implacably hostile. Strangely, however, it is precisely the hostility of the coast, which seems to suggest better things in the interior: errant birds, lost in the thick fog, are emissaries of tolerable and habitable land in the interior. This optimism was shared and even surpassed by his Settle's shipmate, George Best, who imagined a "riche and beneficial" country beyond the coastal mountains.

The following year, in Frobisher's third voyage, Settle found the coast of Friezland as imposing as ever. Frizeland, he writes, is a "hie and cragged land . . . almost cleane covered with snow," where all one can see are "huge hilles . . . all covered with foggie mistes." Even approaching this uninviting land is hazardous, owing to "the great Isles of yce lying on the seas, like mountaines, some small, some big, of sundry kinds of shapes, and such a number of them, that wee could not come neere the shore for them."[34]

As in the previous year, the natural obstacles along the hostile shore appear insuperable. Together, the two descriptions offer a complete list of obstacles that stand in the way of a traveler seeking to penetrate the inviting land beyond. Ice bulwarks, fogs, icebergs, and high snowy mountains are all arrayed against the explorer, creating a continuous, seemingly impenetrable, coastline.

In certain places in his account Settle discards some of these obstacles while adding others. Remarking on the approach to Meta Incognita, for example, Settle warned that "Whoso maketh navigations to those Countreys, hath not onely extreme winds, and furious seas to encounter withall, but also many monstrous and great islands of Yce."[35] On the final approach to Frobisher Bay, Settle predictably observes "so much yce, that we thought it impossible to get into the Straights."[36] Storms, icebergs, and pack ice provide a somewhat different array of obstacles than Settle's 1577 account of Friezland. The unfortunate results for the explorer are, however, the same, as the land is made unapproachable and impenetrable.

Best's description of Friezland follows the same outline as Settle's account. In the 1576 voyage, Friezland was described as "an high and ragged land" which Frobisher "durst not approch . . . by reason of the great store of ice that lay alongst the coast, and the great mists that troubled them not a little."[37] A year later "Friselande sheweth a ragged and high lande, having the mountaines almost covered over with snow alongst the coast full of drift yce, and seemeth almost inaccessible."[38] Even when Frobisher did eventually manage to reach land, he was forced to return quickly to his ships before they disappeared "by the suddaine fall of mistes."[39]

The theme is repeated endlessly in Best's descriptions of the approach to

Frobisher's strait: on Frobisher's first voyage, "there lay so great store of ice all the coast along so thicke together, that hardly his boat could passse unto the shore."[40] The following year the explorers found their passage difficult "by reason of the abundance of yce which lay alongst the coast so thicke togither, that hardly any passage through them might be discovered."[41] In 1578 the *Anne Francis*, after losing her way in the "mistaken strait" (later named Hudson Strait) for several days, found the entrance to Frobisher's strait extremely hazardous, "for yet the yce choked up the passage, and would not suffer them to enter."[42] John Janes, who took part in Davis's expedition in 1586 similarly complains of "mightie mass of yce . . . finding it a mightie barre to our purpose,"[43] to give but a few examples.

Mists were also repeatedly emphasized as part of the defenses of the land. Lost in the mistaken strait for some days, part of the fleet decided to turn back because "by reason of the darke mistes they could not discern the dangers, if by chance any rocke or broken ground should lie of the place, as commonly in these parts it doth."[44] Such, indeed, was the case when Frobisher himself tried to reenter his bay, only to be repulsed "by reason of the darke fogge amongst a number of Ilands and broken ground that lye off this coast, that many of the shippes came over the top of rockes." The most complete version of the "obstacles" theme at the entrance to the golden land is reserved for the third and most difficult voyage: "passing thorow great quantity of yce, by night were entered somewhat within the Streites, perceiving no way to passe further in, the whole place being frozen over from the one side to the other, and as it were with many walles, mountaines, and bulwarks of yce, choked up the passage and denied us entrance."[45] The resemblance to Settle's description of Friezland is unmistakable: walls, mountains, and bulwarks bar the way and prevent access to the interior. The natural fortress is as formidable in Frobisher's strait as it is in Friezland.

There is, of course, a major difference between the theme of the "golden interior" and the element of "obstacles" in the exploration accounts. With hindsight, we can confidently say that Baffin Island was no golden land, and that the supposed gold ore contained nothing of value for Frobisher and his men.[46] We can, therefore, comfortably view this tale as "fiction," a product of the expectations of the explorers rather than of the newly discovered land. The gold was invented, not discovered: it was a pure product of the master narrative of exploration and serves as evidence for its power to produce reality. The case is different with regard to the travelers' reports on the obstacles lying in their way. Those were, undoubtedly, very real and will be encountered by any traveler to the Canadian arctic today, much as they were in the sixteenth century. Nothing in an analysis of the texts will alter the fact that pack ice often clings to the coast of Greenland, or that mists are a serious navigational hazard in those waters.

In discussing Settle's, Best's, and Davis's reports of their experiences as elements of a narrative, I am certainly not claiming that they were as imaginary as the gold of Meta Incognita. My point is, rather, that the ice shelf, mists, icebergs, and mountains encountered by Frobisher's and Davis's expeditions were interpreted and then recounted according to the predetermined formula of the standard tale of exploration. A clear example of this can be seen in Settle's description of Friezland. On encountering the coastal ice shelf, Settle immediately viewed it as a "bulwarke" that "defendeth the countrey." The mountains, the thick fog, and the floating ice were also viewed as obstacles to be penetrated by the general in his attempts to break through the defenses. Although the difficult environment was undoubtedly very real, Settle's view of it as a challenge to be overcome is far from self-evident and is, in fact, rather puzzling. It is particularly surprising if we consider that Friezland was not one of the designated targets of the expedition, and there was no reason to assume that it offered valuable commodities. Indeed, four days of sailing along the coast in 1577 revealed no signs of habitation. But even in the face of such discouraging signs, Settle proved optimistic: on the basis of the most flimsy evidence—the appearance of little birds—he concluded that the interior was tolerable and habitable.[47] His view of the land, his evaluation of its prospects, and his reading of the signs were all guided by his preconceived vision of what a new land is like and how its discovery should proceed.[48]

In his brief description of Friezland Settle manages to produce the main elements of the master narrative of exploration: a promising land in the interior protected by seemingly impregnable natural defenses, and a foreign traveler from the outside attempting to break in. This is indeed an impressive achievement on Settle's part, especially since his actual experience of the land hardly seemed to support his optimistic contentions. Best's and Settle's accounts of the obstacles in the entrance to Frobisher's strait were similarly structured by the same tales. The ice shelf, the fog, and the icebergs were, of course, perfectly real. Their positioning in the accounts as defenses to be penetrated in order to reach a golden land beyond, however, points to their role in a preconceived narrative of exploration.

This brings us to the third and most important element in this narrative— the discovery of passages. Like the "riches of the land" motif, the "discovery of passages" theme operates both in the search for the Northwest Passage and in Frobisher's local activities in Meta Incognita. The original goal of Frobisher's and Davis's voyages was the discovery of a passage to Cathay, and in Davis's case it remained unambiguously the sole purpose of his voyages. When the narrative of exploration is applied to this search, the entire continent of America is perceived as a barrier that must be traversed in order to proceed to the South Sea and the wealth of the Orient. When Frobisher replaced the riches of Cathay with the golden ore of Meta Incognita, however, "the discovery of passages" took on a new meaning: instead of seeking to sail right through the continent, Frobisher now searched for passages to take him past the natural fortifications of the coast and into the golden interior of the land. The "passages through obstacles" theme, like the "land and its riches," were thus displaced from their original setting, only to reappear on a local scale in the unlikely setting of Baffin Island.

Settle's first description of Frobisher Bay provides an excellent illustration of the original use of the narrative, the one concerned with the passage to Cathay. His account of Frobisher's attitude to the discovery of the bay is revealing:

> [On] the 16 of the same [July], we came with the making of land, which land our Generall the yeere before had named the Queenes foreland, being an Island as we judge, lying neere the supposed continent with America: and on the other side, opposite to the same, one other Island called Halles Isle, after the name of the Master of the ship, neere adjacent to the firme land, supposed continent with Asia. Betweene the which two Islands there is a large entrance or streight, called Frobisher's streight, after the name of our Generall, the first finder thereof. This said streight is supposed to have passage into the sea of Sur, which I leave unknowen as yet.[49]

Frobisher required remarkably little evidence in order to proclaim the bay he had discovered to be the long-sought passage to the Orient. The break in the coastline appeared to him to be a "large entrance" to a passage to the South Sea. It is certainly no coincidence that the two islands positioned symmetrically at the mouth of the strait in Settle's account are so reminiscent of the Pillars of Hercules. Like them, they mark the end of the known world and the entrance to a new and uncharted one. With completely unwarranted confidence Frobisher declares one side of the opening to be "America," the other "Asia." Even Settle seems concerned about these interpretations and prefers to talk about the "supposed continent" and "supposed passage" that are still unknown. Frobisher, however, is confidently reading the familiar narrative of exploration onto the natural features of the alien shore.

Best confirms Settle's tale of the first discovery of Frobisher's strait in all its essentials:

> And sailing more Northerly alongst that coast, he descried another forland with a great gut, bay or passage, divided as it were two maine lands or con-

tinents assunder . . . wherefore he determined to make proof of this place, to see how farre that gut had continuance, and whether he might carry himselfe thorow the same into some open sea on the back side, whereof he conceived no small hope . . . and that land upon his right hand as he sailed Westward he judged to be the continent of Asia, and there to be divided from the firme of America, which lieth upon the left hand over against the same.

Best then adds a further dimension to the account, by comparing Frobisher's achievement to Magellan's discovery of the southern straits half a century before: "This place he named after his name, Frobishers Streights, like as Magellanus at ye Southwest, having discovered the passage to the South sea (where America is divided from the continent of that land which lieth under the South pole) and called the same straights, Magellanes straits."[50] The fictional source of Frobisher's claims is made more apparent here than in Settle's account. Frobisher's confidence that he is at the entrance to the Northwest Passage is based on an analogy with Magellan's discovery. The southern passage is here used as a model for what Frobisher can expect to find in the north. The familiar elements are present in this description of the Straits of Magellan: the continuous bulwark of land barring the way to the open seas to the west, and the narrow open passage that threads its way between the land masses. Our familiar story is at work here, and Frobisher transfers it wholesale to the north in interpreting his own discoveries.[51]

The master narrative of exploration has here produced its first tangible geographical result: a long continuous strait stretches from Frobisher Bay in the east, through the continent of North America and into the South Sea. Such, at least, was the view taken by Frobisher and his fellow explorers. With this discovery, their expectations regarding the Northwest Passage seemed to have been justified.

"Frobisher's strait" was not to be the only contribution of the master narrative of exploration to northwestern geography. It seems, rather, that the tale of discovery was endowed with an extraordinary capability to produce and reproduce straits and passages wherever it was applied. Two more passages to the Orient were "discovered" by Frobisher and Davis when they applied this narrative to the search for the Northwest Passage. First of these was Frobisher's mistaken strait, which he entered by accident on his third voyage. After sailing west through the strait for several days, he reluctantly turned back and joined the rest of his fleet in Frobisher Bay, all the while proclaiming to his men that "if it had not been for the charge and care he had of the Fleete and fraughted ships, he both would and could have gone through to the South Sea, called Mar del Sur, and dissolved the long doubt of the passage which we seek to find to the rich countrey of Cataya."

According to Frobisher, the further they traveled in the mistaken straight, "the wider we found it, with great likelihood of endlesse continuance. And where in other places we were much troubled with yce, as in the entrance of the same, so after we had sayled fiftie or sixtie leagues therein we had no let of yce, or other thing at all, as in other places we found."[52]

Two significant points in Frobisher's account should be noted here: first is Frobisher's unyielding confidence that any break in the continuous coast is his promised strait to the riches of Cathay. The same tale that previously led him to insist that Frobisher Bay is the entrance to the Northwest Passage, now convinces him that the same is true of the mistaken strait. The master narrative of exploration has created its second passage to the Orient. The other significant issue here is that the main elements of this tale are repeated on a smaller scale in the description of the passage itself. The strait moves from difficult icy waters at its entrance, to open waters further on, with the promise of "endlesse continuance" from there. The story is familiar by now: the same role played by the coastal ice-shelf and mountains of Friezland, or the mists and icebergs of Frobisher Bay, is here assigned to the difficult entrance to the straits. The inevitable promised land of the interior takes on the character of a free expanse of water open to sailing. Indeed, in John Davis's account the promise of clear waters beyond treacherous bulwarks of ice was a recurrent theme in the accounts of the northwest explorers.

The third major passage to the Orient was "discovered" by John Davis in 1585. While sailing along the west coast of Greenland, Davis recounts how his fleet came upon what he calls "our first entrance to the discovery:

> The weather being very foggy, we coasted this North land; at length when it brake up, we perceived that we were shot into a very faire entrance or passage, being in some places twenty leagues broad, and in some thirty, altogether void of any pester of ice, the weather very tolerable and the water the very colour, nature and quality of the maine ocean, which gave us greater hope of our passage.[53]

The main theme here is perfectly familiar by now. The "entrance to discovery" is hampered by ice and obscured by mists. Beyond these obstacles lies a clear unobstructed passage in good weather, which is revealed once the fog disperses.

Davis is even more confident in his *Worldes Hydrographical Description*, published after his return in 1595. There he notes that "neither was there any yce towards the North, but a great sea, free, large very salt and blew, & of an unsearchable depth." He then concludes, "By this last discovery it seemed most manifest that the passage was free & without impediment toward the

North."[54] The similarities with the descriptions of Frobisher Bay and the mistaken strait in Settle's and Best's accounts are obvious. Once more, with the aid of the familiar narrative of exploration, a passage has been created that breaks through the American land mass and Arctic ice and leads to the golden lands of the East.

Such were the geographical manifestations of the master narrative of exploration when applied to the search for the Northwest Passage. Long corridors were thus established, passing through the mainland into open seas and rich lands beyond. It has already been noted, however, that this was not the only use made of this narrative by the explorers. Frobisher, after all, spent most of his time looking for riches in Meta Incognita rather than in Cathay. The same tale proved just as useful in this context as it was in the search for the Northwest Passage.

It seems, in fact, that much of Frobisher's activities in Meta Incognita consisted of finding passages and penetrating through walls of ice to open space. Examples abound: "At our first comming," Settle writes,

> the streights seemed to be shut up with a long mure of yce, which gave no litle cause of discomfort unto us all: but our Generall, (to whose diligence imminent dangers, and difficult attempts seemed nothing, in respect of his willing mind, for the commoditie of his Prince and Countrey,) with two little Pinnesses prepared of purpose, passed twice thorow them to the East shore, and the Islands thereunto adjacent.[55]

Settle's account of the next voyage is almost identical: "The second of July . . . we fell with the Queenes foreland, where we saw so much yce, that we thought it unpossible to get into the Straights: yet at last we gave the adventure and entred the yce."[56] And Settle continues, "At the first entring into the yce in the mouth of the Straights, our passage was very narrow, and difficult but being once gotten in we had a faire open place without any yce for the most part."[57] Best similarly describes the fleet's attempt to enter Frobisher Bay in 1578 as a passage through "mighty mountains."

A somewhat different theme that, nonetheless, closely parallels the "breaking through" trope is the passage from fog to light. Settle describes the dangers to the fleet when "through darknesse and obscuritie of the foggie mist we were almost run on rocks and Islands before we saw them: But God (even miraculously) opening the fogges that we might see clearly."[58] Later on he describes how "the fogges brake up and dispersed, so that we might plainely and clearely behold the pleasant ayre, which so long had bene taken from us, by the obscuritie of the foggie mistes."[59] Best is similarly thankful that "it pleased God to give us a cleare of Sunne and light for a short time to see and avoyde thereby the danger, having bene continually darke before."[60] The passage through dark fog to clear light and through bulwarks of ice into open water are clearly parallel. Both are manifestations of the "passages" theme of the master narrative of exploration when it is applied to Frobisher's activities in the Canadian arctic.

Perhaps the most interesting instances of the use of the "passages" trope occurs at the intersection of the two levels of Frobisher's enterprise—the search for the Northwest Passage, on the one hand, and the mining of gold in Meta Incognita, on the other. This occurred when the fleet was swept into the mistaken strait in the 1578 voyage. The discovery of this entrance posed a considerable geographical problem to Frobisher: since he was already committed to the notion that Frobisher's strait was the eastern entrance to the Northwest Passage, was it then possible that there were two separate passages leading to the South Sea? The notion seemed unlikely, and Frobisher therefore concluded that the two passages must be connected. All that remained was to identify their meeting point, and the fleet set about determining just that. In one place Best recounts, "Our Men that sayled furthest in the same mistaken straights . . . affirme that they met with the outlet or passage of water which commeth through Frobisher's Straights and followeth as all on into this passage."[61]

Later on Best tells a slightly different version of this story: while returning eastward from his venture into the mistaken strait, Frobisher "perceived a great sound to go thorow into Frobisher's Straights." According to Best, Frobisher decided to prove that the two passages do indeed converge at that point: "Whereupon he sent the Gabriel the one and twentieth of July, to prove whether they might go thorow and meete againe with him in the straights, which they did." This, Best acknowledges, has interesting geographical implications for the land lying between the two passages: "as we imagined before," he claims, "so the Queene's foreland prooved an Iland, as I think most of these supposed continents will."[62]

This entire account is, of course, an obvious fallacy. Frobisher's "strait," as any modern map will show, is merely an inlet on the coast of Baffin Island. It has no western entrance, and it nowhere comes close to joining up with Hudson Strait. No ship, then or now, can pass directly from Hudson's to Frobisher's "straits." It is probably no coincidence that Best gives a surprisingly brief account of this exploit, limiting himself to a laconic "which they did." The lack of details is probably a sign of Best's keen awareness that in this case at least he was recounting not what he had observed and experienced on his voyage, but rather what he hoped and aspired to discover.

The problem faced by Frobisher with the discovery of the mistaken strait could be summed up as a crisis of credibility. It arose because the standard narrative of exploration had overextended itself here and created one transcontinental strait too many. Frobisher's strait was already established in his accounts as a passage to the East, and Frobisher had staked his reputation on its viability. The new passage offered equally good, if not better prospects of such a passage and therefore posed a serious dilemma. Frobisher could, in theory, declare that the previous "passage" was a mistake and that he had now discovered the true one. This, however, would severely damage his credibility, for it would expose the fact that his previous assertion that Frobisher's strait was the way to Cathay was based on the flimsiest evidence and could be easily discarded. Barring this, he could claim to have discovered two separate passages to the South Sea, one originating in Frobisher's strait, and one in the mistaken strait. This, however, seemed an equally damaging assertion, for what was the likelihood that there existed two distinct transcontinental passages, running parallel courses through North America? Such a claim would again bring upon Frobisher the suspicion that he was prone to discover "passages" where none existed, aided more by wishful thinking and an active imagination than actual observation.

Faced with these two unattractive options, Frobisher concluded that the only acceptable explanation for the existence of two passages to the Orient was that they were, in fact, one and the same. The two passages had to be linked, and Frobisher again utilized the standard tale of exploration in establishing the connections between them. The inherent capability of the tale of exploration to carve passages through uncharted lands, was now used to solve the problem of the proliferation of pathways to the Orient. New (and fictional) local passageways were created through the "Queen's Foreland," the land mass separating Frobisher Bay from Hudson Strait, and a ship, the *Gabriel*, was sent on an imaginary voyage to confirm their existence. In the process the land was again dissected, and new islands were formed. Best himself extended the process even further and quickly concluded that what was true locally would be true of the land as a whole. Just as "the Queene's foreland prooved an Iland," he wrote, so will "most of these supposed continents."[63] It is no wonder that in the maps and accounts of Frobisher's expedition the North American landmass is transformed into broken land and carved up by islands and passages. This unique geography was created by the repeated application of the master narrative of exploration to the lands of the Canadian arctic.

The three themes of the master narrative operate as a unit in their assault on the physical continuity of the new lands. First riches are posited in the interior, to serve as a mythical goal for the expedition. Seemingly insurmountable obstacles are then placed around them, and those in turn succumb to the persistent efforts of the enterprising explorers, who locate passages through them. The chroniclers of Frobisher's and Davis's voyages utilize this basic story over and over again on all levels of their accounts, with the result that the land they are describing is broken down into islands, straits, and passages.

A poem by Abraham Fleming, dedicated to Frobisher, prefaces the first edition of Dionise Settle's account of the voyages. Equating Frobisher to Ulysses and Jason, it captures the essence of the tale of the adventurous traveler who breaks through all obstacles to reach the golden fleece.

> Through sundrie foming fretes, and storming streightes,
> That ventrous knight of Ithac's soyle did saile:
> Against the force of Syrens, baulmed beightes,
> His noble skill and courage did prevaile.
> His hap was hard, his hope yet nothing fraile.
>   Not ragged Rockes, not sinking Syrtes or sands,
>   His stoutnesse staide, from viewing foreign lands.
>   . . .
> A right Heroicall heart of Britanne blood
> Vlysses match in skill and Martiall might:
> For Princes fame, and countries speciall good,
> Through brackish seas (where Neptune reignes by right)
> Hath safely sailed, in perils great despight:
>   The Golden fleece (like Iason) hath he got,
>   And rich returned, saunce losse or lucklesse lot.[64]

THE ENGLISHMEN ARRIVE IN VIRGINIA

The search for a passage to the Orient dominated the efforts of English explorers in the late 1570s. Frobisher's search for the Northwest Passage exactly coincided with Drake's voyage of circumnavigation through the Strait of Magellan. While Davis continued the search (although on a much reduced scale) in the following decade, the main English imperialist enterprise of the 1580s was Walter Raleigh's attempt to establish a permanent colony in North America.

The two enterprises, the search for a passage and the colonization of the land, were undoubtedly intertwined. When Frobisher's expedition shifted its focus from searching for passages to Cathay to digging for gold on Baffin Island, Meta Incognita was transformed from an unfortunate obstacle on the way to the East into a land worth possessing. Frobisher therefore fully intended to leave a permanent settlement on Meta Incognita in his third voyage, to keep watch over the mines and continue the search for the elusive passage. Only the destruction at sea of the prefabricated structure, which was to serve as the colony's home, prevented him from implementing this plan.[65] Conversely, although Raleigh's intent was clearly to establish a permanent colony in Virginia, his project nonetheless retained the discovery of a passage to the Orient as a supplementary goal.[66]

The tragic story of Raleigh's Virginia colony is well known.[67] In 1584 Raleigh obtained a letter of patent from Queen Elizabeth, granting him the right to explore and settle "such remote, heathen and barbarous lands, countries, and territories, not actually possessed of any Christian prince."[68] That same year he sent a scouting expedition, headed by Philip Amadas and Arthur Barlowe, to survey the land north of Florida and select an appropriate location for settlement. The two returned from their voyage with wondrous tales about the riches of the land and recommended the island of Roanoke, off the coast of modern North Carolina, as an appropriate site for settlement. The new land, known to the Indians as "Wingandaoca," was renamed Virginia in honor of Queen Elizabeth.

The following year Raleigh organized a colonizing expedition under the command of Ralph Lane. The settlers landed on Roanoke in June of 1585 and remained in Virginia a full year, during which time they mounted several surveying expeditions of the region. The mathematician Thomas Hariot and the painter John White took an active part in these expeditions and produced a record of the land and its inhabitants unique for the period.[69] By the summer of 1586 supplies had run out, relations with the local Indians had deteriorated drastically, and the promised fleet, bearing supplies from England, had not arrived. A fleet commanded by Francis Drake stopped at the settlement in June and attempted to supply it with a ship and provisions. When the designated vessel was lost in a summer storm, the settlers abandoned the colony and took passage with Drake back to England. The relieving fleet arrived at Roanoke a few weeks later, finding only the vacant remains of the settlement.

Immediately upon the return of the colonists, Raleigh renewed his efforts to settle Virginia. The colonists, with John White as governor this time, left England in May of 1587 and landed on Roanoke in late July. White stayed for only a month, returning to England with the colonizing fleet to organize a supporting expedition with fresh supplies. By the time White's supporting fleet was ready, however, England was threatened by a Spanish invasion and all ships were ordered to remain in port. Despite White's persistent efforts, only in 1590 did he manage to return to Roanoke to search for the colonists. He did not find them: the Roanoke settlement was deserted, and clues left by the settlers indicated that they were headed for the island of Croatoan. Bad weather prevented White from landing there, and he was forced to turn back to England, never again to return to America. The fate of Raleigh's lost colony remains unknown.

Our interest in the Roanoke enterprise is much the same as it was in the case of the northwest voyages. The focus is not so much on the details of the events themselves, but on the rhetoric of exploration and its basic narrative, which shaped the explorers' understanding of their task and their view of the new land. The master narrative of exploration is present in Raleigh's colonizing project much as it was in the voyages of Frobisher and Davis. Riches are posited in the interior and various obstacles bar the way to them, only to be dissected by passages located by the determined explorers. In the process, the geography of Virginia is reconfigured to meet the demands of the tale: the hostile coast is broken down and penetrated by a series of bays, sounds, and rivers, all potentially leading to the rich interior.

The report of Amadas and Barlowe serves a clearly defined role in this narrative: it posits amazing riches in the interior, worthy of being sought by the explorers and settlers who will follow in their footsteps. Upon first landing on the Carolina Banks, they report,

> wee viewed the lande about us, being whereas we first landed, very sandie, and lowe towards the water side, but so full of grapes, as the very beating, and surge of the sea overflowed them, of which we founde such plentie, as well there, as in all places else, both on the sande, and on the greene soile of the hils, as in the plaines, as welle on every little shrubbe, as also climbing towardes the toppes of the high Cedars, that I think in all the world the like aboundance is not to be founde: and my selfe having seene those partes of Europe that most abound, finde such difference, as were incredible to be written.[70]

The tone of wonder and amazement at the fertility of the land continues when they describe a local fisherman who "in lesse than halfe an howre . . . had laden his boate as deepe as it could swimme."[71] They confidently conclude, "The soile is the most plentifull, sweete, fruitfull, and wholesome of all the world."[72]

Ralph Lane, governor of the 1585 colony, was quite as enthusiastic in his communications to the queen's secretary, Sir Francis Walsingham, while in Roanoke: "All the kingedomes and states of Chrystendom theyere commodytyes joyegned in one together, doo not yealde either more good, or more plentyfulle whatsoever for publyck vse ys needefull, or pleasinge for delyghte."[73] In another letter he suggests that "if Virginia but had Horses and Kine in some reasonable proportion, I dare assure myself being inhabited

with English, no realme in Christendome were comparable to it."[74] It is probably no coincidence that both these letters were written in the first two
months of Lane's stay in Virginia. Even considering the injunction of the master narrative of exploration to posit great wealth and riches in the land, it is
doubtful whether Lane's exuberant optimism would have survived the adver-
sity that the colony faced later on.

Thomas Hariot's *A Briefe and True Report of the New Found Land of
Virginia* is a very different text from Amadas and Barlowe's idealized account
or Lane's optimistic reports to Walsingham.[75] Although it was written shortly after Hariot's return from Virginia and aimed at raising money and volunteers for further expeditions, it is a far more serious and detailed text than
its predecessors. Most of it is composed of a detailed analysis of the commodities that the country could "yield" to industrious Englishmen. It does not
display the unbridled optimism and superlative language that characterizes
the other texts. Despite its restrained tone, however, the *Report* does posit
great riches in the land and invites the readers to explore and colonize it.

The *Report* is first and foremost a list of valuable commodities that can
be obtained in the new land: grass silk, worm silk, flax and hemp, allum,
wapeih, pitch, tar, rozen and turpentine, sassafrass, cedar, wine, oil, and
many more items are listed and discussed in succession, with a single concise
paragraph allotted to each. This arrangement has a double effect: on the surface, it simply recounts in detail the various advantages that could be gained
by settling the land. The subdued, factual tone is important here, because it
grants credibility to the *Report* and urges the reader to accept it as truth. On
a deeper level, it is not so much the substance of Hariot's account that promotes the riches of Virginia, but rather the form of presentation. The
arrangement of the commodities in a long monotonous list, which continues
predictably page after page, obscures the distinct qualities of the various
items discussed, but stresses the overall abundance. In the end, the tract is not
promoting the particular qualities of the separate commodities, but the
incredible richness of the land, reproduced in the seemingly endless list. It creates an image of a land overflowing with valuable commodities, of which the
settlers can simply take their pick. The rhetoric of Hariot's tract is, no doubt,
different from the tone of Amadas and Barlowe's account. The overall
image projected, however, is rather similar in both cases.

Another aspect of *A Briefe and True Report* relevant in this context is
Hariot's belief that further in the interior even greater riches lie hidden. The
settlement of Virginia would be worthwhile, he writes in the conclusion, even
"if there were no more knowen than I have mentioned." All this, however,
"doubtlesse is nothing to that which remaineth to bee discouered:

> For although all which I have before spoken of, have bene discovered & experimented not far from the sea coast ... yet sometimes we made our iourneies farther into the maine and countrey; we found the soyle to bee fatter; the trees greater and to grow thinner; the grounde more firme and deeper moulde; more and larger champions; finer grasse and as good as ever we saw any in England; in some places rockie and farre more high and hillie ground; more plentie of their fruites; more abundance of beastes; the more inhabited with people, and of greater pollicie & larger dominions, with greater townes and houses.[76]

Since the interior is so much more promising than the coastal area, he concludes, "Why may we not then looke for in good hope from the inner parts of more and greater plentie, as well of other things, as of those which we have already discovered?"[77]

The notion that further in the interior there are ever greater riches awaiting discovery seems to be a fixture of the master narrative of exploration. We have already encountered it in Settle's and Best's optimistic descriptions of Friezland and Meta Incognita. Hariot's basis for his claim was as shaky as theirs, for it relied on nothing more than the tale of exploration itself. The myth of discovery, it seems, supplied Hariot with a general map of Virginia, just as it supplied Frobisher's chroniclers with the outlines of their understanding of Baffin Island.

Hariot's displacement of the mythical land from the coastal regions to the interior serves an apologetic function as well. Amadas and Barlowe had described the entire land of Virginia as an enchanted land beyond the ocean to their audience in England, and Hariot attempted to support this vision by conveying an impression of an incredibly bountiful land in his list of commodities.

Hariot's position was, nonetheless, quite different from that of his predecessors: by the time he wrote his *Report* he had already lived in Virginia for a full year, explored and studied it in detail, and suffered hunger and hardships with the other colonists. Despite the positive tone of his description, the magical quality of the land, so prominent in Amadas and Barlowe's account, had been sadly diminished. A list of commodities, however long and impressive, is also finite, and its limits are as obvious as the abundance it represents. A land endowed with such commodities is impressively bountiful, but it is no match to the truly magical land suggested by the exploration narrative. Hariot's solution to this problem is much like Frobisher's, when faced with the possibility that the "mistaken strait" rather than Frobisher's strait is the true passage to the East. Rather than relinquish the tale of exploration, Hariot chooses to replicate it: the true magical land undoubtedly exists, according to Hariot, but it is located further inland than Amadas and

Barlowe had suspected. By thus replicating the narrative and pushing the golden land to the interior, Hariot manages to preserve the basic outlines of the master narrative.

The belief that Virginia holds great riches in the interior, lying in wait for enterprising explorers and settlers, was expressed repeatedly by those involved in the enterprise.[78] In most cases this richness was described in terms of the amazing fertility of the land, which was the main theme in the descriptions of both Hariot and Amadas and Barlowe.

The presence of precious metals is also mentioned occasionally: Hariot refers to rumors that "farther into the countrey" there are "Rivers that yeelde also white graynes of Mettall, which is to be deemed Silver."[79] Elsewhere Hakluyt the elder refers confidently to "the mines there of Golde, of Silver, Copper, Yron, etc."[80] Now if the imperialists' view of the agricultural and commercial potential of Virginia was greatly exaggerated, their belief in the presence of precious metals was almost certainly groundless. Yet it seems that no undiscovered land was complete without gold and silver lying in wait in its entrails. Gold and silver came to serve as an abstract sign for the desirability of the land and the valuable secrets it holds.

Agricultural commodities, even if incredibly abundant, did not seem to the promoters of the enterprise to be sufficient inducement for a colonizing enterprise. Only the presence of gold and silver could project the image of the new land as possessing great and hidden riches. Only precious metals could produce the necessary crusading zeal among potential adventurers and draw them to explore the Arcadian landscape of America. By locating "gold" in Meta Incognita, Frobisher transformed the arctic wastes of Baffin Island into a promising substitute for the rich kingdoms of the Orient. By a similar move, Hakluyt and Hariot hoped to configure Virginia as a golden land awaiting its discoverer.[81]

Once the land has been established as rich and desirable, the narrative moves on to deal with the difficulties obstructing the way to the interior, and the passages through them. A trace of this theme can be found in Amadas and Barlowe's account, even though their report was concerned more with observing great riches than possessing them. While sailing along the North American coast on July 2, 1584, they report, "We found shole water, which smelt so sweetely, and was so strong a smell, as if we had bene in the midst of some delicate garden, abounding with all kinds of odoriferous flowers, by which we were assured, that the land could not be farre distant." Two days later,

> We arrived upon the coast, which we supposed to be a continent, and firme lande, and we sailed along the same, a hundred and twentie English miles, before wee could finde any entrance, or river issuing into the Sea. The first

that appeared unto us wee entred, though not without some difficultie, and cast anker about three harquebushot within the havens mouth, on the left hande of the same.[82]

The main components of our tale are all present here: the wondrous land in the interior, compared to a "delicate garden"; the protective bulwark, so similar to those described by Frobisher's chroniclers, which guards the land and prevents easy entry by the explorers; and finally the inevitable passage through which the traveler enters the land, "though not without some difficultie."

John White charts out this tale in his maps of Virginia, which were based on the expeditions he undertook with Thomas Hariot during their stay in Roanoke. Appropriately, they were published as part of Hariot's *Briefe and True Report* in Theodor De Bry's edition of 1590.[83] In a map entitled "The arrival of the Englishemen in Virginia" White depicts all these elements (see Figure 3). The coastal barrier with its irregular breaks stretches from one side of the map to the other. Beyond it awaits rich wooded Virginia, sparsely populated by the local Indians, who can be seen hunting and fishing while the corn grows unattended in the fields. Outside the banks the English fleet lies at anchor, facing toward the land, seeking to penetrate the coastal bulwark. The hazards of such an attempt are indicated by sunken ships, which lie at the entrance to each and every one of the apparent passages. A small boat has, nonetheless, managed to break through the obstacles and is approaching the village of Roanoke bearing the sign of the cross.

Hariot's comments on White's map are revealing:

> The sea coasts of Virginia arre full of Ilands, wherby the entrance into the mayne land is hard to finde. For although they be separated with divers and sundrie large Division, which seeme to yeelde convenient entrance, yet to our great perill we proved that they were shallowe, and full of dangerous flatts, and could neuer perce opp into the mayne land, untill we made trialls in many places, with or small pinnes. At lengthe wee fownd an entrance uppon our mens diligent serche therof . . . saylinge further, wee came to a Good bigg yland.[84]

Significantly, as David Quinn points out, Hariot's description here of the landing in Virginia does not fit any of the accounts of either the 1584 or the 1585 expeditions.[85] Quite as significantly, it does bear a striking resemblance to the coastal descriptions of Frobisher's expedition. For what Hariot and White offer here is not a precise account of the actual landing in Virginia, but a paradigmatic tale of what a landing should always be. A rich land, coastal

Figure 3. John White, "The Arrival of the Englishemen in Virginia." Printed in Thomas Hariot, *A Briefe and True Report of the New Found Land of Virginia*, in Theodor de Bry, ed., *America*, vol. 1 (Frankfurt am Main, 1590).

barriers, and passages between them are simply the necessary components required by the narrative that structures the description of the landing.

White's general map of Virginia, entitled "Americae pars, Nunc Virginia dicta" completes the picture (see Figure 4). Again we have an English fleet hovering outside the banks, while the local Indians fish and hunt inside its protective shell. As before, the breaks in the sea wall are marked as shallow and hazardous. This map, however, extends much further into the interior and thus introduces a new dimension into the story. While the "Arrival" map focused exclusively on the breaks in the offshore barrier, the general map extends this theme to passages from the coastline to the interior. The inner coastline is riddled with wide sounds and rivers, which seem to offer easy passage to the heart of the land. Some of these are merely local streams that lead nowhere. Some appear in the map to lead further inland, but their true source remains unknown. Most promising of all is a wide river at the center of the map, which leads deep into a mountain range and possibly beyond. It is prob-

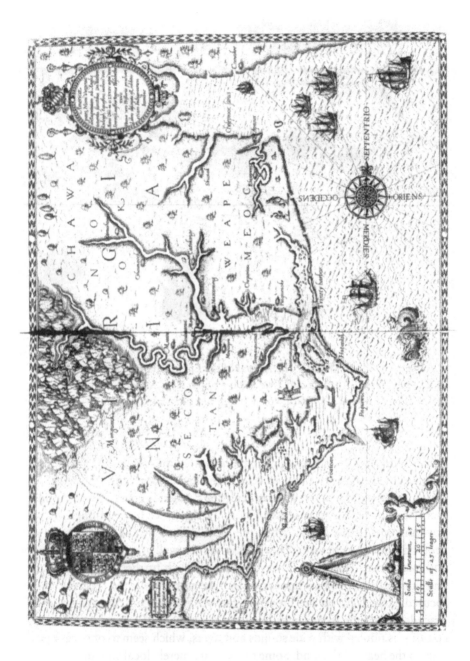

ably no coincidence that Hariot located the source of the precious metals of Virginia in just such a mountainous region in the interior. The message is clear: wide and open passages lead from the shoreline to the rich interior. Hariot recreated Amadas and Barlowe's magical land, moving it further inland. White follows this move by supplementing the breaks in the coastal banks with deeper passages leading to this hidden country.[86]

White's maps tell a tale, but they are not works of fiction. Like the fog and ice of the Canadian arctic, the Carolina Banks are obstinately real and do not simply accommodate any story that a cartographer may choose to tell about them. The coastal barriers in White's maps are, therefore, both "real," in the sense that they represent actual conditions along the Carolina coast, and "mythical," in the sense that they convey an abstract general truth about exploration in general. In this double identity they are true to Barthes's notion of myth, in which wide-ranging ideological claims are presented, cloaked as specific and transparent "facts." The exactitude and reliability of White's maps serve to enhance their power as instantiations of the imperial tale of exploration. They are the standard narrative of exploration in cartographic form.[87]

The double function of White's maps is made clearer by examining a map from a different context, which is far less "realistic" than White's maps, and yet looks remarkably like them. The map in question was published in Venice in 1565 in the third volume of Giovanni Baptista Ramusio's *Navigationi et Viaggi* and depicts the island of "Norumbega" (see Figure 5). While the name "Norumbega" usually referred to the area of New England in this period, Ramusio's map depicts the island of Newfoundland and the lands lying opposite it across the gulf of St. Lawrence. Much like Frobisher's treatment of Baffin Island, Ramusio carved up Newfoundland and the American coastline into an archipelago, with straits and rivers providing passage between the various islands. This in itself may alert us to the possibility that a narrative of exploration is in operation. Even more significant are the obvious similarities of the map to White's depictions of Virginia: a long continuous sand-bank protects Norumbega from the outlying ocean, just as the Carolina Banks do in White's Virginia. A French ship is shown hovering outside the barrier, while smaller crafts managed to break through and enter the inland sea beyond and plant a cross on one the islands. The Indians of

Figure 4 (opposite).   John White's map of Virginia, 1590. Printed in Thomas Hariot, *A Briefe and True Report of the New Found Land of Virginia*, in Theodor de Bry, ed., *America*, vol. 1 (Frankfurt am Main, 1590).

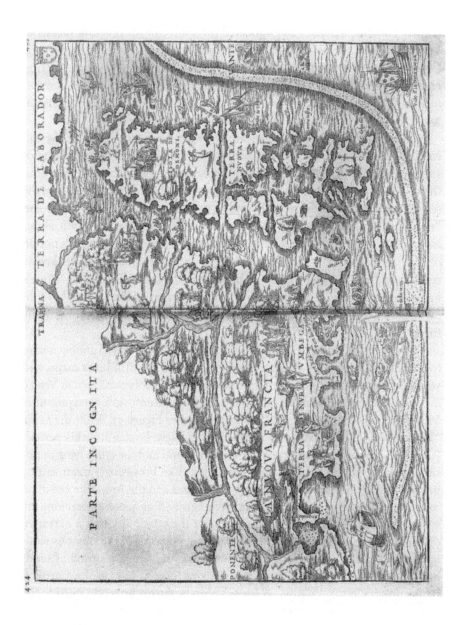

Norumbega, like those of Virginia, are shown fishing and hunting, quite oblivious to the intruders who now menace their bountiful land. Finally, wide rivers lead from the coastline into a mountainous and unexplored land further in the interior.

White's maps are geographically far superior to Ramusio's chart. The general outlines of the Carolina coast, as depicted by White, are easily recognizable in any modern map of the region. The same cannot be said of the Norumbega map, with its plethora of nonexistent islands and coastal passageways. But precisely because of its geographical inaccuracy, the ideological narrative of the map is foregrounded in a way that is not possible in White's more careful depictions. Since the actual geographical characteristics of the land do not serve as much of a constraint on Ramusio's depiction of Norumbega, it is much easier to identify the features of the map as ideological claims, elements of a tale of discovery. The bountiful land of the interior, protected by coastal barriers but penetrated by deep passages, clearly composes an ideological tale here rather than a careful geographical depiction. Since White's maps are much more faithful to the geographical features of the land than Ramusio's, it is more difficult to pinpoint their mythical characteristics and separate them from the unyielding contours of the Carolina coast. Their striking similarity to Ramusio's fantastic chart of a completely different region helps foreground the founding myth that structures them both. The geography of White's Virginia, much like Ramusio's Norumbega, is shaped by the standard narrative of exploration.

THE TREASURES OF EL DORADO

In his lengthy report on the 1585 Virginia colony, Ralph Lane recounted a conversation he had with an Indian king named Menatonon, whom he held in captivity for a few days. Menatonon greatly impressed Lane, who describes him as "a man impotent in his lims, but otherwise for a savage, a very grave and wise man, and of very singular good discourse." Equally impressive was the tale he related:

> Amongst other things he told me that going three dayes iourney in a canoa vp his riuer of Choanoke, and then descending to the land, you are within foure dayes iourney to passe ouer land Northeast to a certaine King's country, whose

Figure 5 (opposite).   Giovanni Battista Ramusio, Map of Norumbega. Printed in Giovanni Battista Ramusio, *Delle Navigationi et Viaggi*, vol. 3 (Venice, 1565) (Courtesy of the Huntington Library).

province lyeth vpon the Sea, but his place of greatest strength is an Iland situate as he described vnto me in a Bay, the water round about the Iland very deepe.

Out of this Bay, hee signified vntoo mee, that this King had so great quantitie of Pearle, and doeth so ordinarily take the same, as that not only his own skins that he weareth, and the better sort of his gentlemen and followers, are full set with the sayd Pearle, but also his beds, and houses are garnished with them, and that he hath such quantitie of them, that it is a wonder to see.[88]

The name of the country, we later learn, was Chaunis Temoatan. Encouraged by this report, Lane mounted a search to locate this land. The expedition nearly ended in disaster when Lane and his men ran out of supplies and were attacked by Indians on the way. On the verge of starvation, they managed in the end to fight their way back to their base on Roanoke.[89]

The search for Chaunis Temoatan was only a minor episode in Raleigh's Virginia voyages of the 1580s. Far more time and effort was spent in attempts to establish a permanent colony than in the pursuit of mysterious kingdoms. Lane's search was, nonetheless, a harbinger of things to come: ten years later, Raleigh was engaged in a large-scale enterprise in search of a hidden land that bears an unmistakable resemblance to Chaunis Temoatan. This was, of course, the kingdom of Manoa, more famously known as El Dorado.

The kingdom of Manoa, according to legend, was founded by refugees from the empire of the Incas when it succumbed to Francisco Pizarro's conquistadors in the 1540s. Led by Tupac Amaru, a member of the royal house, they set out into the South American heartland, eventually settling in a great plain on the shores of a lake. There they reestablished an empire of wealth and splendor, on a scale that far surpassed its predecessor. Its ruler was known as the gilded man, or El Dorado in Spanish, and the land itself was often referred to by that name. During the course of the sixteenth century, many adventurers set out to find and conquer this lost empire of the Incas. The most famous attempt was the disastrous 1559 expedition led initially by Don Pedro de Ursua, and later by Lope de Aguirre. In the 1580s and 1590s the search was taken up by Antonio de Berrio, who after traversing the continent several times, determined that the empire must be located in the highlands of Guyana.[90]

Walter Raleigh, in all probability, first heard of the lost kingdom in 1586 from Don Pedro Sarmiento de Gamboa, the Spanish colonizer of the Straits of Magellan, whom he briefly held captive.[91] In 1594 Raleigh sent his lieutenant Jacob Whiddon on a preliminary survey to the coast of Guyana, and in February of the following year he set out on a major expedition himself. The story of his voyage is narrated in full in the account he published after his return, entitled *The Discoverie of the large and bewtiful Empire of*

Guiana.[92] Raleigh's first stop was the island of Trinidad, off the coast of Guyana, where Berrio was at the time waiting for reinforcements to renew his quest for El Dorado. The English sacked and burned the town of San Josef, capturing Berrio and his second in command. After interrogating his prisoners, Raleigh set out toward the mainland and advanced slowly up the Orinoco. He reached as far inland as the village of Morequito, where the river is joined by one of its tributaries, the Caroni.

Throughout the voyage Raleigh tried consistently to establish friendly relations with local Indians. His purpose, he explains in his account, was to secure their help in future operations against the Spaniards and, if necessary, against the kingdom of Manoa. This policy seemed to pay off when he encountered the Indian king Topiawari of Arromaia, whom he described as "our chiefest friend."[93] Topiawari, like Menatonon a decade earlier, provided the English with the story they expected: Many years ago, he recounted, there came "a nation from so far off as the *Sun* slept . . . with so great a multitude as they could not be numbred or resisted . . . they wore large coats, and hats of crimson colour." This western people defeated the local Guyanans, and built "a great town called *Macureguarai* at the said mountaine foote, at the beginning of the great plaines of *Guiana*, which have no end."[94] Raleigh, optimistic as ever, took this to mean that El Dorado lay just beyond the nearest mountain range. His attempts to reach it, however, were frustrated by the difficult navigation in the river and the summer rains. He therefore promised Topiawari that he would return the following year and sailed back to England, convinced that he was on the threshold of a great discovery.

Raleigh's reception in England fell far short of the enthusiasm he had hoped for. Despite the quick publication of *The Discoverie of Guiana*, skepticism persisted among potential backers of the enterprise. Raleigh, therefore, had to settle for a very small expedition in 1596 led by his lieutenant, Lawrence Keymis. As he was given only two small ships to command, Keymis never really considered carrying out the conquest of Guyana envisioned in the *Discoverie*. Finding Raleigh's passage of the previous year blocked by a new settlement established by Berrio (who had been released by Raleigh the previous year), Keymis conducted a close survey of the coast of Guyana in search of alternative passages to the interior. He found no less than forty potential coastal entrances, many of which, he believed, led directly to the great lake on the plain beyond the mountains. Although the Orinoco was now blocked by the Spanish fort, he reported on his return to England, other, better passages existed which led to the golden kingdom of Manoa.[95]

The story of El Dorado, and Raleigh's search for it, can be viewed as the paradigmatic case of the master narrative of exploration. The kingdom of Manoa possesses all the attributes posited by the general tale: an incredibly rich land on the shores of a lake, surrounded by mountains, seemingly unapproachable. Raleigh and Keymis, furthermore, took it upon themselves to act their own part in the tale: they were the enterprising explorers who would find the passages through the land's defenses and penetrate to the golden interior. Raleigh attempted this by sailing up the Orinoco, the major waterway of the region. Keymis's method followed the classical move of the standard narrative of exploration: he systematically searched for rivers along the coast, which would break the continuity of the land and lead to its hidden heart. When all these elements are combined, they make for a perfect example of the narrative and its operation—from the setting up of the geographical presupposition to the carving up of the land that results from it.

Not only the completeness of the elements of the narrative but also their abstract, nonspecific character point to the paradigmatic nature of the story of El Dorado. The very name given to the land, "El Dorado," seems to suggest this. "Manoa," or "Macureguarai" are specific place names, much like Lane's "Chaunis Temoatan" or Frobisher's "Cathay." El Dorado, however, is an abstract notion: it refers not to a geographical location, but to the specific quality of the land that is central to the story—its richness in gold. The fact that El Dorado is not a place name at all is further evidenced by the search for it, which ranged all over South America. What kind of a place is it which could be practically anywhere? The only thing that mattered about the land of El Dorado is that it fit the parameters required by its myth; its actual geographical location and characteristics were irrelevant.

Perhaps most crucial to the abstract nature of the story of El Dorado is the fact that it was, of course, completely imaginary. Frobisher's geography of the Canadian arctic was the result of a clash between the master narrative, which he brought ready-made from England, and the unyielding realities of the American Northeast. Similarly, White's depictions of the coast of Virginia represented the actual contours of the land as much as the general tale of exploration. The kingdom of Manoa, however, simply never existed. There was no second empire of the Incas in Guyana or anywhere else. El Dorado was, therefore, a straightforward expression of the story of exploration, operating in complete freedom from geographical constraints.[96] It was the pure, abstract, and paradigmatic version of the standard narrative.

The guiding presence of the tale is evident throughout the texts associated with the enterprise. Raleigh openly associated the search for El Dorado with the mythical riches of Peru and the East when he claimed it to be "very likely" that "the Emperour *Inga* hath built and erected as magnificent pal-laces in *Guiana*, as his auncestors did in *Peru*, which were for their richness and rareness most marveilous and exceding al in *Europe*, and I think the world." Raleigh then proceeded to describe the approaches to the land, claiming that "those that are desirous to discover and to see many nations, may be satisfied within this river, which bringeth forth so many armes and branches, leading to severall countries, and provinces, above 2000 miles east and west, and 800 miles south and north. And of these, the most eyther rich in Gold, or in other merchandizes."[97]

Keymis's participation in the master narrative is evident when he encounters his own Indian chief, who predictably supplied him with yet another version of Topiawari's (and Menatonon's) tale: "By this Captaine I learned that Muchikeri is the name of the country where *Macureguerai* the first town of the Empire of *Guiana*, that lyeth towardes *Raleana*, is seated in a faire and exceeding large plaine, belowe the high mountaines that beare Northwesterly from it, that it is but three dayes iorney distant from *Carapana* his porte: and that *Monoa* is but six daies farther."[98] "Raleana" was the name given by the English to the Orinoco. The chief then proceeded to detail the various other rivers that provide passage to this fair land.

Positioning the hidden land, however, is not enough. One must also induce fellow adventurers to join in its conquest. Keymis, ever aware of the shaky support for the voyages, decried his countrymen's incredulity and passivity. If Englishmen will not take advantage of the miraculous opportunity presented to them, he warned, others undoubtedly will: "In a second age, when in time truth shall have credite, and men woondering at the richnesse, and strength of this place, which nature her selfe hath marveilouslie fortified, as her chiefe treasure house, shal mourn and sigh and hold idle cickles, whilest others reap & gather in this harvest."[99] In his appeal, Keymis here brings to the fore the two central elements of the master narrative. The land itself is, as usual, posited as rich and fortified, and this is presented as a challenge to English manhood to break through the fortifications. Both the promised land and the enterprising explorer are posited here by Keymis, and the tension between them enhanced by the failure of the English to take on the challenge.

In other places both Raleigh and Keymis utilize overt sexual metaphors to make their point. Raleigh, in the final and most famous passage from the *Discoverie*, describes Guyana as a virgin land awaiting her conqueror:

> To conclude, *Guiana* is a Countrey that hath yet her Maydenhead, never sackt turned nor wrought, the face of the earth hath not beene torne, nor the vertue and salt of the soyle spent by manurance, the graves have not beene opened for gold, the mines not opened with sledges, nor their Images puld down out of their temples. It hath never been entred by any armie of strength, and never

Conquered or possessed by any Christian Prince. It is besides so defensible, that if two fortes be builded in one of the provinces which I have seen . . . no shippe can passe up.[100]

Raleigh is here caught in a balancing act between the need to promote Guyana as a well-defended land, in accordance with the master narrative, and the fear such a description might discourage potential adventurers. To fend off the obvious implication that it would be difficult to gain possession of a country so well protected, he adds rather implausibly that "there is . . . great difference betweene the easines of the conquest of *Guiana*, and the defence of it being conquered."[101]

Keymis uses a similar metaphor in a similar context. While Englishmen waste their energies trying to defraud their neighbors of a petty "halfe an akor of land," in Guyana, he assures the reader, "whole shyeres of fruitfull riche groundes lying now waste for want of people, do prostitute themselves unto us, like a faire and beautiful woman, in the pride and flower of desired yeares."[102]

By gendering the land as female, Raleigh and Keymis balance the apparently contradictory themes of their description: the land is virginal, closed, and protected, while at the same time it must inevitably open up to the enterprising explorer. If the land is presented as a woman in a sexual encounter, seemingly impenetrable but ultimately willing, then the contradiction can be maintained.[103] The sexual metaphor serves another purpose in providing the necessary connection between the hidden land and "her" explorer. The adventurers are provided with the well-defined role of seducers, who can consciously rob and despoil Guyana and yet view their actions as the ultimate fulfillment desired by the land itself. It is no coincidence that Raleigh and Keymis both turn to this gendered language when addressing their skeptical countrymen. If Guyana is a beautiful woman, then the difficulties in approaching her can be seen as challenges, which enhance her desirability, rather than as disheartening obstacles. The eventual opening of the passage, by diplomacy or by force is, furthermore, assured.[104]

All of this brings us back to the theme of passages to the interior. Raleigh's search can be viewed as a headstrong pursuit of the single great passage to Manoa, namely, the Orinoco River. His reference to "so many armes and branches leading to several countries" hints at the potential of the exploration discourse to multiply itself and carve up the land. His main focus, however, is on the great river, which he named "Raleana" after himself. In contrast, Keymis in 1596 searched systematically for all possible passages to the interior and dissected the Guyanan coastline in the process. Indeed, parts of his narrative are simply lists of river openings along the coast, which, he promises, "do all fal out of the plaines of the Empire."[105]

While sailing along the coast, Keymis describes a small stretch of land in which "there fall into the sea these severall great rivers *Arrowari, Iwaripoco, Maipari, Coanawini, Caipurogh.*" The next day the fleet encounters the "*Arcooa, Wiapoco, Wanari, Caparwacka, Cawo, Caiane, Wia, Macuria, Cawroor, Curassawini.*"[106] In another place he lists the rivers of "*Cuna-namma, Uracco, Mawari, Mawarparo, Amonna, Marawini, Oncowi, Wiawiami, Aramatappo, Camaiwini, Shurinama, Shurama, Cupanamma, Inana, Curitini, Winitwari, Berbice, Wapari, Maicaiwini, Mahawaica, Wap-pari, Lemerare, Dessekebe, Caopui, Pawarooma, Moruga, Waini, Barima, Amacur, Aratoori, Raleana.*"[107] All of these rivers empty into the Atlantic along the Guyanan coast, and Keymis fully expects them to lead him inland to the golden kingdom. He summarizes his description by announcing to the queen:

> This your second discoverie hath not onlie found a free & open entrance into *Raleana*, which the Naturals call *Orenoque*: but moreover yieldeth choise of fourtie severall great rivers (the lesser I do not recken) being for the most part with small vessels navigable, for our merchants and others. . . . To such as shall be willing to adventure in search of them, I could propose some hope of gold mines, & certain assurances of peeces of made golde, of Spleen-stones, Kidney-stones, and others of better estimate.[108]

Gold and riches in the interior, coastal breaks and passages on the exterior: the operations of the exploration discourse in structuring Keymis's account are unmistakable.[109]

As was the case with Frobisher's expedition and the Roanoke voyages, the Guyana exploration narrative is best captured in visual form. In the British Library there is indeed a map of Guyana associated with Raleigh's enterprise that was probably prepared by Thomas Hariot (see Figure 6).[110] The map captures all the elements of the master narrative of exploration: the center is squarely occupied by the large "Lake of Manoa," which lies upon the wide "Valley of Guiana." Several towns are depicted in the valley, along the shores of the lake and close by. A continuous mountain range separates this "plane of the empire" from the coastal region to the east. The shoreline itself is broken by a continuous series of river openings, though only two actually break through to the valley—the "Orenoke" to the north and the Amazon to the south.

In this map, Manoa is indeed a natural fortress, marked by a lake and surrounded by mountains. It is reachable, nonetheless, through the deep and

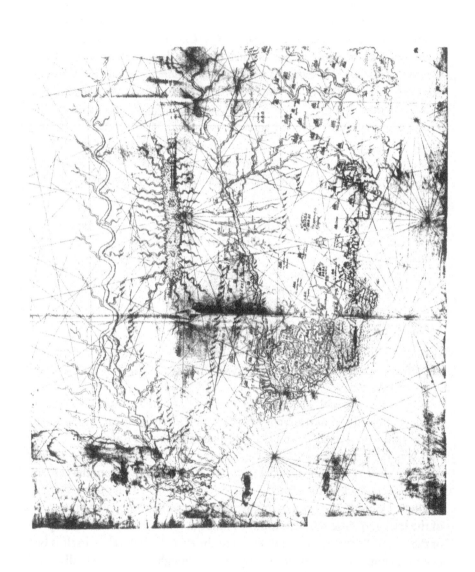

navigable rivers, which penetrate to the interior all the way to its hidden center. The land of Guyana is thus carved up by a long series of rivers and streams, which break its coastline at regular intervals. Like the story of El Dorado, the map is an expression of the standard tale of exploration in its purest, most abstract form.[111] In Guyana, as in Roanoke and Meta Incognita, the narrative helped structure the geographical contours of the land.

Keymis dedicated a Latin poem to Hariot at the beginning of his *Second Voyage*, describing his friend as "Matheseos, & Universae Philosophiae peritissimum." The poetic qualities of Keymis's rhymes are questionable, but his poem succinctly sums up the rhetoric of the enterprise of Guyana, which permeates its narratives, maps, and documents.

> Surrounded by many mountains as if by walls this land;
> Around it flow the waters of Raleana.
> Guiana has deep bounteous entrances;
> · · ·
> This land has gold, and gems the color of grass.
> It is always spring there; there the prodigal earth yearly
> Teems, it puts down the sun in fertility.
> · · ·
> There we explore or nowhere. Therefore we ask;
> god grant that we possess this Canaan. Amen.[112]

HENRY BRIGGS AND THE RIVER PASSAGE

The English search for the Northwest Passage, which occupied Frobisher and Davis in the 1570s and 1580s, fell into a lull in the later decades of the sixteenth century. The war with Spain, as well as the Virginia and Guyana enterprises, occupied the energies of the English magnates of exploration and eclipsed their earlier attempts to locate an alternative route to the Orient. The accession of James I to the throne in 1603 brought about a renewed interest in locating the elusive passage. James was far more careful to avoid offending Spanish interests than his predecessor and tried to avoid challenging their claims to sovereignty in America. The Guyana project was greatly toned down, and the colonization of Virginia was handed over to private commercial interests.[113] The safest way to promote English expansion under the

Figure 6 (opposite).   Thomas Hariot(?), map of Guyana, ca. 1596. British Library Add. MS 17940. Note that the map is pointed southward (Courtesy of the British Library).

new policy was to focus on areas as far away as possible from the Spanish dominions. Since Spain clearly considered the Straits of Magellan to be within her domains, English expeditions once again concentrated on finding a passage to the East through North America.

Between 1602 and 1631 no less than ten different English expeditions set out in search of the Northwest Passage (as well as several searching for a northeastern one.) Henry Hudson (1610–11), Thomas Button (1612–13), Robert Bylot (1615, 1616), and Luke Foxe (1631) are only the most famous seamen involved in the intense effort to locate the elusive passage during this period.[114] In the context of this renewed interest in the Northwest Passage we come across the final text to be discussed here—"A Treatise of the northwest passage to the South Sea, through the Continent of Virginia and by Fretum Hudson," by the Savilian professor of mathematics, Henry Briggs.[115]

It is curious to note that Henry Briggs was Thomas Hariot's exact contemporary, the two having been born in 1561 and 1560, respectively. But while both were considered among the leading mathematicians of their time, the career paths they followed differed markedly. Hariot went the way of personal patronage, becoming a member of the household of Sir Walter Raleigh and later of the Earl of Northumberland. In contrast, Briggs belonged to more public institutions: in 1592 he was appointed a reader of medicine at Cambridge, and in 1596 he became the first professor of geometry at Gresham College. In 1620, he was invited by Sir Henry Saville to take over his duties in Oxford and became the first Savilian professor of mathematics. The parallel careers of Briggs and Hariot epitomize the two main routes of advancement open to mathematicians at the time.[116]

Briggs was a strong advocate of geographical exploration and was much occupied with problems of mathematical navigation. Following the suggestion of his friend William Gilbert, Briggs believed that the vertical dip of the magnetic needle at a given location on the globe provided an indication of latitude.[117] Gilbert's supposed discovery was potentially of immense value to navigators, since it would free them from their dependence on celestial observation in determining latitude. Briggs proceeded to construct elaborate mathematical tables, which were to be used by mariners in determining their location according to the magnetic dip. The tables were published in 1602 as part of Thomas Blundeville's *The Theoriques of the Seven Planets*, and reprinted in 1613 in Marke Ridley's *A Short Treatise of Magneticall Bodies and Motions*.[118]

In the following years Briggs cooperated closely with Edward Wright, who established his reputation in maritime circles in 1599 by publishing the first tables of meridional parts, necessary for constructing a Mercator projection map.[119] In 1610 Briggs supplied several navigational tables to the second edition of Wright's *Certaine Errors of Navigation*, and in 1616 he joined forces with him in publishing the English translation of John Napier's book on logarithms. As the introduction to the translation makes clear, navigation was considered the primary field of application for Napier's new technique.[120] Briggs devoted most of his remaining years to the systematic calculation and publication of logarithmic tables.

Considering Briggs's continued involvement in navigational issues, it is hardly surprising to find him as an advisor and promoter of the 1631 voyages in search of the Northwest Passage. Two competing expeditions were set to sail in that year, led respectively by Luke Foxe and Thomas James. Before embarking on their voyages both captains sought the old mathematician's support for their venture. Foxe described Briggs as a "former acquaintance" who deserved credit for granting him the opportunity to embark on his expedition. James was, apparently, not previously acquainted with Briggs, but he considered the mathematician's blessing so vital that he made a special trip from Bristol to Oxford to ask the Savilian professor to acknowledge his expedition's priority over Foxe (Briggs reportedly refused).[121] The captains' rivalry for Briggs's patronage continued into the voyages, and both named geographical discoveries after their mathematical patron. James boasted a "Briggs' Bay," while Foxe assigned a group of islands in Hudson Bay the ungainly name "Briggs, His Mathematickes."[122]

Briggs's prominence in the Northwest Passage enterprise stemmed, no doubt, from his 1625 "Treatise of the Northwest Passage to the South Sea, through the Continent of Virginia and by Fretum Hudson."[123] The tract is an enthusiastic endorsement of the existence of a passage to the East through North America. As the title itself makes clear, the exact route is relatively unimportant: the passage may be found either through the land of Virginia or through Hudson Strait or, perhaps, through both.[124] It is not clear from the title—or from the text itself—whether Briggs is arguing for the existence of two separate passages, or whether the two are, in fact, connected and form a single passage. It is even possible that he is simply claiming that at least one of the two passages exists. The important point for Briggs is that a passage is undoubtedly there, awaiting its discoverers.

The vagueness shrouding Briggs's claims in the treatise seems rather surprising at first. After all, Briggs is making a geographical argument, and such arguments draw their authority precisely from their specificity. Tide patterns through a strait, for example, may indicate the existence of a large body of water beyond; a great river flowing into the ocean may indicate the existence of a large land mass in the interior, and so on. In each case the argument is based on the specifics of a geographical location, and conclusions are drawn from its unique characteristics. Briggs's ambivalence about the exact nature of his claims points to a different type of argument altogether. The geographical particularities of North America, it seems, do not matter much to Briggs. The passage may be by way of rivers, it may be through an oceanic strait, or it may be through both. The existence of the passage is assumed independently of the geographical realities and is known before any exploration takes place. Future geographical discoveries will eventually fill in the actual details of the Northwest Passage. The truth of its existence is known beforehand.

Since Briggs's confidence in the Northwest Passage is clearly not based on detailed geographical considerations, where does it come from? Briggs is simply following the inner logic of the standard narrative of exploration. Indeed, our familiar narrative has left its heavy mark throughout the "Treatise of the Northwest Passage." Briggs begins by describing Virginia as a wondrous land of plenty, much in the manner of Hariot's *Briefe and True Report*. "The ayre is healthfull and free both from immoderate heate, and from extreame cold; so that both the inhabitants and their cattell do prosper exceedingly in stature and strength, and all Plants brought from any other remote climate, doe there grow and fructifie in as good or better manner, then in the soyle from whence they came."[125]

Briggs goes on to praise the climate, the humidity, and the excellent location of the land, lying between Europe and the West Indies.[126] Showing an impressive spirit of international cooperation, Briggs invites voyagers of other nations to rest and recuperate in Virginia, "whether they be of our owne nation, or our neighbours and friends."[127] He continues, "The multitude of great and navigable Rivers, and of safe and spacious harbours, as it were inviting all Nations to entertaine mutuall friendship, and to participate of those blessings which God out of the abundance of his rich Treasures, hath so gratiously bestowed some upon these parts of Europe."[128] Briggs's peaceful rhetoric is very different from the belligerent pronouncements of his Elizabethan predecessors and is, undoubtedly, politically motivated. The only hope for promoting American exploration under the early Stuarts lay in insisting that such activities would not be perceived as hostile to Spanish interests in America. Briggs is, therefore, eager to portray the colonization of Virginia as a peaceful enterprise, promoting harmony and cooperation rather than rivalry among nations. The "neighbours and friends" who would refresh themselves in Virginia on their voyages between Europe and the West Indies are, of course, the Spaniards, and Briggs insists that the colony would be beneficial even to them.

More interesting to us than the imperial politics of Briggs's presentation is the standardized vision he offers of the land itself. Indeed, Briggs's description of Virginia is clearly in the tradition set by Amadas, Barlowe, and Hariot in the 1580s. For Briggs, as for his predecessors, Virginia's climate is ideal, the land immensely fertile, the harbors sheltering, and the rivers inviting. When he concludes that Virginia has been endowed by God with many "rich treasures," he is merely summarizing his account of the wealth of the land. It is a wondrous land of plenty, the storehouse of many of God's wonders.

Briggs's depiction of Virginia already brings to mind some major components of the standard narrative of exploration. A desirable and wondrous land of riches is, after all, the primary condition of the tale, which sets the exploration project in motion. The geography of the land may also begin to seem familiar by now: Briggs's emphasis on the navigable rivers that penetrate the land brings to mind the straits and passages that characterized the Canadian arctic in the accounts of Frobisher's voyages. Keymis's depictions of the "deep bounteous entrances" of Guyana, and his painstaking listing of the numerous rivers of the land are even closer in content and style to Briggs's Virginia. A wondrous land of riches, dissected by deep passages, is indeed a hallmark of the geography set by the standard narrative of exploration.

There is also a seemingly major difference between Briggs's account and the exploration geography posited by his predecessors. For Frobisher and Keymis the riches of the land lay deep in the interior, and the rivers served as passages leading to that hidden wealth. In Keymis's case, the rivers led through mountains to the plains of the empire of Manoa; in Frobisher's case, the many straits offer passage to the rich Orient or the gold deposits of Meta Incognita. Briggs, by contrast, makes no precise claims to the location of the riches of Virginia, and it is not clear initially whether the rivers lead to them at all. Although the geography of Briggs's Virginia is indeed reminiscent of Guyana and Meta Incognita, the precise workings of the narrative of exploration are not clear at first glance. What, it may be asked, is the function of the "great and navigable rivers" in Briggs's account, and how, if at all, do they fit within our familiar tale?

The answer becomes clear once we note that the wealth of Virginia itself is a sideshow in Briggs's treatise. The true importance of Virginia is that it provides a route to the riches of the East. Even more important than its placement as a hub of Atlantic trade, Briggs argues, is its location "in respect of the *Indian* Ocean, which wee commonly call the South Sea." This ocean, he promises, "lieth on the West and Northwest side of Virginia, on the other side of the Mountains, beyond our Fals, and openeth a free and faire passage, not onely to *China, Iapan,* and the *Moluccas;* but also to those rich Countries of Terra *Australis,* not as yet fully discovered."[129]

Virginia here is not a golden land itself, but rather a barrier and a passage on the way to the Orient. A clear and open route to the East lies just beyond reach, in the interior of the land. All an explorer needs to do is break through the mountains, pass beyond the falls, and he will be well on his way to China, Japan, and other undiscovered rich lands.

The role of the navigable rivers in Briggs's tale is now clarified: "If we shape our iourney towards the Northwest following the Rivers towards the head, we shall undoubtedly come to the Mountains, which as they send divers great Rivers Southward, into our Bay of Chesepiock, so likewise doe they send others from their further side Northwestward into that Bay where *Hudson* did winter."[130] The rivers, then, provide passage to the interior of the land, through the mountains and beyond, opening the way to the golden kingdoms of the Orient.

The parallels with Raleigh's and Keymis's Guyana are unmistakable. In Briggs's Virginia, as in Keymis's Guyana, a mountain range is posited in the interior, blocking the passage from the coast to the rich lands beyond. In both cases, great and navigable rivers break through the mountains and provide access from the coast to the golden lands to the West. All an explorer needs to do, according to both Keymis and Briggs, is follow the rivers to their sources, and his way will then be open to the treasures of El Dorado, China, or Japan.

The similarities to Frobisher's Meta Incognita are striking as well. Frobisher, like Briggs, was looking for a passage through the continent of North America to the great civilizations of the East. His conviction that such a passage existed tended to transform every inlet he encountered into a likely passage leading to the Pacific Ocean. The inlet now known as Frobisher Bay was to him "Frobisher Strait," and he confidently identified the northern side of the passage as a promontory of Asia. When, on a later voyage, he strayed into "Hudson's Straits," he duly identified it as another "northwest passage." In order to limit the plethora of "passages" thus created, Frobisher was forced to conclude that all his distinct "straits" were connected and were, in reality, only one passage.

The same dynamic is at work in Briggs's treatise. On the one hand, Briggs is excited about the possibilities opened up by the discovery of Hudson Bay, and he is eager to maintain it as a likely northwest passage. On the other hand, his primary purpose in the treatise is to promote the rivers of Virginia as possible routes to the East. He therefore suggests that "it is not unlikely that the Westerne Sea . . . commeth much neerer" to the rivers of Virginia than Hudson Bay, thus providing a "river passage" to the Pacific Ocean.[131] Briggs, like Frobisher before him, has fallen victim to the tendency of the standard narrative of exploration to produce and reproduce passages wherever it is applied. His solution to the problem of multiple passages is also the same as Frobisher's: some rivers, he posits, flow from the mountains of Virginia "Northwestward into that Bay where Hudson did winter."[132] The rivers of Virginia and Hudson's strait, then, are not two distinct passages at all, but merely separate entrances to the same passage. Briggs, like Frobisher before him, attempts to control the number of passages produced by the tale of exploration by drawing connections between them.

At the end of the treatise Briggs makes use of one more commonplace of the standard tale of exploration—the reports by the natives of the wealth that lies waiting in the interior. Ralph Lane drew encouragement from King Menatonon's tale of Chaunis Temoatan. Raleigh relied on Topiawari's testimony on the kingdom of Manoa, and Keymis described an encounter with an Indian "Captaine" who supplied him with information on the exact topography of the kingdom. Since Briggs was not an explorer, he could not supply the firsthand accounts of his predecessors. This shortcoming, however, did not deter him from utilizing this popular resource of the tale of exploration. His contention that the South Sea is easily reachable from Virginia, he claims, "is much confirmed by the constant report of the Savages, not onely of Virginia, but also of *Florida* and *Canada*." All of these, dwelling in different places, agree "in the report of a large Sea to the Westwards . . . where they describe great Ships not unlike to ours." This independent testimony of so many different peoples strongly suggests to Briggs "that our endevours this way shall by God's blessing have a prosperous and happy successe."[133] Briggs sums up the accounts of all the Menatonons and Topiawaris of the eastern seaboard of North America into a single "report of the Savages." Their promise to Briggs, as it was also to Lane, Raleigh, and Keymis, was that lands of great riches await their discoverer further in the interior.

The standard narrative of exploration has a major formative role in structuring Briggs's tract. Its presence is evident in Briggs's preconceived conviction of the existence of the passage through Virginia, independently of any particular geographical characteristics of the land. It is evident in the account of Virginia itself as a land of wealth and wonder, in terms reminiscent of Amadas and Barlowe's report. It is evident as well in the unmistakable similarities between the "Treatise" and the earlier accounts of the voyages of Frobisher, Raleigh, and Keymis. But most of all, the master narrative of exploration leaves its mark on the actual geography of the land that Briggs describes. Not far to the west, according to Briggs, lies the South Sea, with

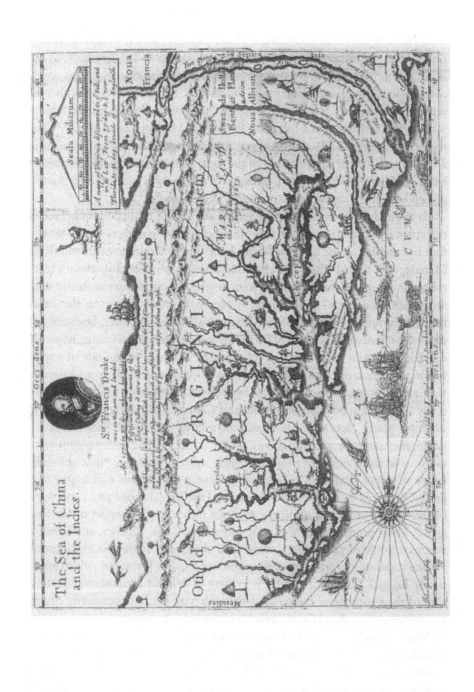

its promise of the riches of China, Japan, and the Moluccas. The coast of Virginia is separated from this sea of promise by a mountain range (presumably the Appalachians) that runs along the coast and blocks access to the West. Fortunately, many great and navigable rivers break through this mountainous barrier. The enterprising explorer in search of the Northwest Passage must simply follow the rivers to their sources, and his way will then be open to the wondrous riches of the East.

As a final note it is worthwhile to observe a map of Virginia, published in 1651 by John Ferrar (see Figure 7). Despite the fact that Ferrar's map saw light a full twenty-six years after Briggs's treatise, it is nonetheless structured according to the basic principles of Briggs's narrative. Virginia, in Ferrar's depiction, is shown as a narrow strip of land dividing "Mare Atlanticum" from "The Sea of China and the Indies." The only obstacle separating the eastern coast of North America from the South Sea is a single mountain range, which runs parallel to the shores. Fortunately for the explorers, who hover in their ships off the Atlantic coast, wide rivers penetrate from the eastern seaboard deep into the mountain range. Once there, Ferrar promises, a mere "ten day march" along "rich adjacent valleys, beautified with as profitable rivers which necessarily must run into ye peaceful Indian Sea," will complete the voyage to Pacific Ocean.

Ferrar's map is clearly structured by the same vision of Virginia as Briggs's "Treatise." The land of riches to the west, the continuous barrier, and the passages that break through are all present in Ferrar's map, just as they were in Briggs's account. Even the explorers seeking passage, who were the target audience of Briggs's "Treatise," are depicted in Ferrar's map in their seagoing ships. Ferrar, in other words, follows Briggs and maps out the geography of the standard tale of exploration upon the land of Virginia. In 1651 John Ferrar was simply the latest (and probably one of the last) to view America strictly in terms of the standard narrative of exploration. His map was a late link in a long tradition. Martin Frobisher's chroniclers, Thomas Hariot's *Report* on Virginia, Raleigh's and Keymis's accounts of Guyana, and Briggs's "Treatise" on the Northwest Passage had all preceded Ferrar in reconfiguring their target lands to accord with this standard tale. James Beare's map of Frobisher's voyages, John White's map of the Carolina coast, and Thomas Hariot's map of Guyana had preceded Ferrar in giving concrete geographical shape to the

Figure 7 (opposite). John Ferrar, Map of Virginia. Printed in Edward Williams, *Virgo Triumphans: or, Virginia Richly and Truly Valued,* 3d ed. (London, 1651) (Courtesy of the Huntington Library).

familiar story. In all these cases, the physical geography of the target lands was depicted in much the same way: great land masses and formidable obstacles were carved up and dissected by rivers and straits, leading from the coast to golden land beyond. In other words, the geography of the new lands of America was configured by the standard narrative of exploration.

CHAPTER 3

Mathematical Empires

MATHEMATICIANS AND THE CULTURE OF EXPLORATION

Thomas Hariot was an English patriot as well as a mathematician: on the eve of one of Walter Raleigh's voyages in search of El Dorado, Hariot dedicated a poem to his famous patron, which he named "Three Sea Marriadges":

> Three new Marriages here are made
> One of the staffe and the sea astrolabe
> of the sonne and starre is an other
> which now agree like sister and brother
> And charde and compasse which now at bate
> will now agree like master and mate.
>
> If you use them well on this your journey
> They will be the king of Spaine's atarney
> To bring you to silver and Indian gold
> which will keep you in age from hunger and cold
> God speed you well and send you fair weather
> And that agayne we may meet together.[1]

The quality of these verses indicates why Hariot was known to his contemporaries as a brilliant mathematician rather than as a poet. Nonetheless, the main theme of the poem perfectly expresses a fundamental reality that Hariot shared with almost all other mathematical practitioners in Elizabethan England: mathematics, for him and his colleagues, was part and parcel of the imperialist project of the age.

The alliance between mathematics and maritime exploration was apparently struck at the founding moment of the English mathematical tradition. Robert Recorde, often considered the father of English mathematics, was an advisor to the Muscovy Company, the first of the great overseas trading companies of London. Significantly, he dedicated his book on advanced arithmetic, *The Whetstone of Witte*, "to the right worshipfull, the governers, con- silles, and the rest of the commitee of venturers into Moscovia."[2]

Many English mathematicians who followed in Recorde's footsteps showed an even stronger affiliation with the imperialist enterprise: John Dee was famous for his support of English expansion and served as technical advisor to numerous voyages.[3] Edward Wright, who translated John Napier's book on logarithms into English, is probably best known for his work on car- tography. He was the first to calculate and publish "meridional parts" tables, necessary for the construction of charts according to the Mercator projection.[4] Henry Briggs was regularly consulted on matters of navigation and also instituted a lecture series on the subject in Gresham College. Thomas Hariot was Raleigh's principal advisor for his expeditions to Virginia and Guyana throughout his career. His unpublished papers contain several unpublished tracts concerned with shipbuilding, mapmaking, and naviga- tion.[5] This involvement in the exploration project was not limited to the great men of Elizabethan mathematics. The same tendency has been carefully doc- umented for dozens of less prominent practitioners by E. G. R. Taylor.[6] Al- most without exception, Taylor found, Elizabethan mathematical practi- tioners were concerned with aiding the voyages of expansion and settlement.

The mathematicians' responsibilities covered precisely the issues men- tioned in Hariot's poem: they invented and produced navigational instru- ments, such as the staff and astrolabe; they calculated the astronomical tables necessary to interpret the instruments' measurements; and they drew the nav- igational charts used in the expeditions. Some, like Dee, Hariot, and Briggs, were well-known figures whose reputations extended well beyond the imme- diate circle of mathematical practitioners. Others labored in relative obscu- rity, publishing small tracts on particular technical problems.[7] In almost all instances, however, Elizabethan mathematicians were profoundly involved in the exploration ventures of their day. For the mathematical practitioners of Elizabethan England, in other words, mathematics and geographical explo- ration went hand in hand.[8]

The association of mathematics with the voyages of exploration hardly seems surprising to the modern observer. The mathematical sciences have become so crucial to modern technology that it seems all but inevitable that mathematical practitioners should have an important role in such a venture. The great voyages, after all, were marvels of early modern technological achievement. Ships were built sturdier and more seaworthy to withstand long-distance sailing in the open ocean. More elaborate rigging schemes were introduced to accommodate oceanic conditions. New navigational instru- ments were developed and built and older ones improved upon. Navigational charts, previously limited to the coasts of Europe and the Mediterranean, were now expanded to include distant oceans and continents. Consistent map projections were developed to take account of the curvature of the earth in global maps, an effect that could be ignored in traditional local maritime charts.[9]

The end result of many of these developments and technical improvements was the radical reform of navigational practices in the age of exploration. For centuries, European and Mediterranean navigation had been a form of pilotage. It combined local coastal charts with sailing by "dead reckoning" and, since the twelfth century, use of the magnetic compass. Most importantly, it put prime value on the pilots' personal familiarity with prominent coastal landmarks along the way. This, of course, would not do for the much longer voyages that now became common. Open ocean navigation required daily measurements of the height of the sun and the North Star, as well as a reliable estimate of the ship's speed. Sophisticated astronomical tables were then used in order to convert the measurements into a good approximation of the ship's position.[10]

The new navigational instruments and the complex numerical tables required for their use were, as one would expect, the concern of mathematical practitioners. Indeed, the technical issues mentioned in Hariot's poem seem to point to precisely such a role for mathematicians in the imperialist project. It is, however, somewhat surprising to find that the mathematicians' involvement in the voyages went well beyond technical improvements. The very same practitioners, known for their technical innovations in the fields of cartography and navigation, were equally famous for their role as public promoters of the enterprise and as leading participants in the voyages themselves.

John Dee was perhaps the greatest advocate of English expansion in the first half of Elizabeth's reign and is traditionally credited with coining the term "British Empire."[11] In 1577 he published the *General and Rare Memorials Concerning the Perfect Art of Navigation*, which contained a series of tracts designed to promote English imperialism.[12] Edward Wright not only wrote on marine cartography but also took part in the Earl of Cumberland's raid on the Azores in 1589 and wrote a detailed account of the voyage.[13] Henry Briggs was known for his "Treatise of the Northwest Passage to the South Sea," promoting the search for a passage through North America, as well as for his celebrated logarithmic tables.[14] Briggs even left his mark on the North American landscape when explorers Luke Foxe and Thomas James had prominent landmarks in the Canadian arctic named after him and his "mathematickes."[15]

Thomas Hariot was a prominent member of Raleigh's first colony on

Roanoke in 1585. Upon his return he published *A Briefe and True Report of the New Found Land of Virginia*, which he hoped would promote further exploration and settlement of the land.[16] It is worth noting that despite Hariot's many years of work on mathematics and natural philosophy, the *Briefe and True Report* remained Hariot's only published work to his dying day.[17] The mathematicians, it seems, were more than mere technical aides called on to solve particular localized problems, within a project defined and controlled by their patrons. Rather, they were full-fledged participants in the enterprise itself and, on occasion, the driving force behind it.

But while many mathematicians enthusiastically took part in promoting English imperial expansion overseas, they did not all share a uniform notion of what such an enterprise entailed. Different mathematicians attached differing meanings to the imperial enterprise. Many eagerly adopted the standard narrative of exploration and discovery. This simple story centered on a wondrously rich land, hidden behind mighty and seemingly impenetrable obstacles. The enterprising explorer must locate the breaks in the seemingly smooth defenses and find his way to the great riches of the interior. Some mathematical practitioners made use of precisely this rhetoric when they turned their attention to promoting voyages of exploration and discovery. Hariot drew explicitly and implicitly on this story when he discussed the riches of Virginia or hidden land of El Dorado and mapped their coastlines. Briggs argued for the inevitability of a northwest passage through North America by making use of the familiar outlines of the same standard narrative.[18]

Other mathematicians drew on different notions of the meaning and purpose of imperial expansion. For although the standard tale of exploration provided a prevalent and popular mould for Elizabethans' conceptions of empire, there were other competing notions as well. Such, for example, was the vision adopted by the Elizabethan polymath John Dee, who along with Hariot was the best-known and most respected English mathematician of his day. Whereas Hariot and his colleagues promoted voyages of grand adventure and even military conquest, Dee thought the empire must be acquired first and foremost through judicial means, by proving that the newly found territories belonged to England by right.

In Elizabethan England, mathematics and imperialism were closely interwoven. Dee, Hariot, Wright, and their contemporaries lived and worked in the midst of a culture saturated with imperialist practice and imagery. The ways in which they imagined the exploration enterprise and its goals will be the focus of this chapter. Some, Like Dee, adopted a conservative notion of empire and were suspicious of attempts to acquire one through exploration and force of arms. Others, including Hariot and Wright, wholeheartedly adopted the aggressive posture of Raleigh and Drake and supported it in words and deeds.

Most importantly, some mathematicians were not content to describe actual voyages with the rhetoric of exploration. Rather, they quickly came to apply the very same language to their mathematical work as well. In their hands, the quest for a hidden golden land beyond formidable obstacles became a metaphor for the quest for knowledge in general and mathematics in particular.

HARIOT AND DEE ON AN ELIZABETHAN EMPIRE

John Dee was known to his contemporaries as the foremost promoter of exploration and discovery in his time. Born in 1527 during the reign of Henry VIII, his interest in geographical exploration dates from the 1540s. In those years, he traveled through Europe and studied under the leading geographers of his day—Pedro Nuñez, Gerard Mercator, and Gemma Frisius. In 1553 he was recruited as a technical advisor to a voyage sponsored by the Merchant Adventurers of London in search of the Northeast Passage to Cathay. More than two decades later he was consulted by Martin Frobisher and Humphrey Gilbert with regard to their planned voyages to the Arctic in search of a northwest passage. Dee provided Frobisher's sailors with geographical instruction and later drew a map of the Arctic based on their discoveries. His optimism regarding the prospects of the voyages is reflected in the fact that he secured from Gilbert a grant of rights for all discoveries above the 50th parallel. This probably also reflects Gilbert's own skepticism about the enterprise, because it would have secured most of what is now Canada for Dee's personal use. Several years later, in 1580, he again served as a technical consultant on Captains Pett and Jackman's voyage in search of a northeast passage.

In the *General and Rare Memorials* of 1577 Dee argued through historical and legal precedents that Elizabeth possessed title to a vast Atlantic empire. In order to secure those far-flung dominions, Dee advocated the establishment of a "petty navy royall" to patrol the sea routes and protect the queen's interests.[19] In short, much like Thomas Hariot and Henry Briggs in later years, Dee was a scholarly geographer, a technical advisor to the voyages, and a political promoter of expansion and empire.[20]

But although Dee occupied the same intellectual space as Hariot and Briggs after him, his views on the nature and purpose of imperial expansion differed markedly from theirs. Dee's approach is exemplified in his treatise *Britanici Imperii Limites* of 1578(?) in which he argued for Elizabeth's title to a vast array of overseas possessions and urged her to take control of her lost empire.[21] His arguments were essentially legalistic, based on a long list of English voyages and "discoveries," both historical and apocryphal, accompanied by complex dynastic calculations.

He argued, for example, that Iceland and Greenland were part of Elizabeth's inheritance because she was the direct successor of King Arthur. Since Arthur fought and defeated the Danes, he was, according to Dee, the undisputed King of Denmark. When the Danes, centuries later, proceeded to colonize the North Atlantic islands they were merely adding to Arthur's patrimony, which was then passed through the generations to Queen Elizabeth.[22]

Another example is his gingerly treatment of the Spanish claims to dominion in the New World. Concerned that Elizabeth would accept the legitimacy of the 1493 papal grant of territories to the Iberian powers, he argued that the Pope was in no position to grant what was not his. In case this argument would fail to persuade the queen, he then added that in any case the Pope did not mean to divide the entire world, but only the areas between the northernmost and southernmost latitudes of Spain. Finally, in case Elizabeth still felt that the Spanish claims warranted some consideration, Dee presented his ultimate argument: through complex dynastic considerations, Dee "proved" that it was in fact Elizabeth, and not Philip II, who was the true legal sovereign of the kingdom of Castille.[23]

He concluded:

> Of a greate part of the sea Coastes of Atlantis (otherwise called America) next unto us, and of all the iles nere unto the same from Florida Northerly . . . the Tytle Royall and supreme is due, and appropriate unto your most gratious majestie and that partlie *Iure Gentium*, patlie *Iure Civilis*, and partlie *Iure Divino*, No other Prince or Potentate in the whole world being able to alledge therto any clayme the like.[24]

Dee's position was that all those disparate lands belonged to Elizabeth by indisputable universal law.

While Dee supported voyages of exploration, these were not in the strictest sense necessary. He had proven to his own satisfaction that the English monarch was already master of a grand empire through universal legal consideration. This was true even if no English subject actually set foot in those far-flung domains and no English ships ever set out to explore the New World. Ultimately, for Dee, the empire depended not on actual exploration by land and sea, but on the correct interpretation and application of the natural and divine laws of succession.

In his treatise, Dee brought home his claim that disparate parts of the globe lie naturally within Elizabeth's domain by literally mapping them onto her body: "The single little black circle shown on the left hand side of your majesty's throne, represents Cambalu, the chief city of Cathay . . . meanwhile, by a wonderful omen the City of Heaven happens to be located at the middle joint of the index finger which encircles the hilt of your sword. . . . Thirdly, at the right side of your majesty, the coast of Atlantis is pleased to have its place."[25] The queen's empire here is inseparable from her very body. Exploration and conquest are hardly necessary: the distant corners of the globe are already united in Elizabeth's person.

Dee's imperial imagery suggested a quite different notion of exploration and empire from the vision promoted by Hariot and others that espoused the standard narrative of exploration. As a member of Walter Raleigh's household, Hariot could hardly have been overly concerned about the niceties of universal law. Raleigh was known to his supporters as a privateer and to his Spanish enemies simply as a pirate. He dedicated his life to breaking the Iberian stranglehold on the Americas and to diverting the riches flowing from the New World toward England and his own pocket. America, for Raleigh, was a wondrous undiscovered land, harboring great riches and awaiting her manly discoverer. If Elizabeth were to earn a stake in this new land, it would be by the noble deeds of arms of her subjects overseas, not by the tedious work of homebound scholars dusting off arcane dynastic documents.

Hariot's support for these views is implicit in his participation in Raleigh's various ventures of settlement and exploration. It is exemplified in his *Briefe and True Report* and also stated plainly in the short poem with which this chapter began. For Hariot, as for Raleigh, Drake, and their colleagues, the voyages were a grand adventure for brave men. They had nothing to do with the universal law of nations; they had everything to do with courage, the spirit of adventure, and, more often than not, greed.

Perhaps the best way to compare Dee's views with those of Hariot and his circle is to contrast the visual images they provided to support their views. Dee's views are beautifully presented in the frontispiece to his *General and Rare Memorials* of 1577 (see Figure 8). As a presentation of Hariot's vision, I have selected the pictorial map entitled "The Arrival of the Englishemen in Virginia," which has already been discussed in the previous chapter (see Figure 3 in Chapter 2).[26]

The two images obviously belong to two distinct genres: Dee's is a frontispiece, designed to convey the general claims that will be discussed in the text; Hariot's is a map that illustrates his personal account of the settlement of a new land. It is therefore initially surprising to note the remarkable similarity between the two images: both depict fleets of stylized oceangoing ships

Figure 8. John Dee, frontispiece to *General and Rare Memorials Pertayning to the Perfect Arte of Navigation* (London, 1577).

that serve as the bearers of maritime and imperial power. Both fleets are hovering just off the shore of rich and promising lands, broken by a major river leading to the interior. Here, however, the similarities end. Hariot's image is first and foremost a local detailed map of an American locale—the seaboard of what is today North Carolina—drawn by an eyewitness explorer. It is designed to convey the voyagers' immediate experience to the readers who stayed home. Dee's image, by contrast, is purely symbolic—its ocean a universal sea and its land is a place both everywhere and nowhere. It is meant to express not an immediate experience, but rather an abstract universal idea.[27]

A closer analysis of the iconography of the frontispiece reveals the general claim Dee makes. Elizabeth is seated at the helm of a grand ship called *Europa*, sailing to seize the castle of "Occasio" accompanied by *Europa* and the bull. On shore is the kneeling figure of Republica Britannica humbly beseeching the queen to embark on her imperial mission with a "fully equipped expeditionary force." Overhead, lending their support to the enterprise are the divine light of "Jehova," the archangel Michael, and the cosmic elements of the sun, moon, and stars. The entire universe, in other words, is waiting for Elizabeth to take her rightful place as ruler of a great empire. All she has to do is take hold of the "occasion," and a perfect harmonious order will reign in the world.

Hariot's map tells a different story, which is a version of the standard tale of exploration. The Virginian interior is depicted as a rich and desirable land, where Indians fish, hunt, and grow crops aplenty. The road to possession of this wondrous land, however, is fraught with dangers. A long and nearly continuous chain of islets bars the explorers' fleet's passage to the Virginia coastline and attempts to force a passage are extremely risky. This is made abundantly clear by the sunken ships, which mark each possible passage. Ultimately, a small party of explorers did manage to break through the obstacles in a small boat and sails toward Roanoke Island bearing the sign of the cross.[28]

The contrast between the narratives of the two images could hardly be greater. For Dee, empire is part of a general cosmic constellation that prevails everywhere; for Hariot, only actual travel and presence overseas can earn an empire. For Dee, empire is a divine gift presented to Elizabeth, while for Hariot, only human efforts can bring it about. In Dee's scheme the queen merely had to take advantage of the "occasio" and take charge of her extensive lands. In Hariot's story, the empire is won through risks and hazards taken on by enterprising explorers, who break through forbidding obstacles and reach the Promised Land of the interior. In short, whereas for Dee imperial title could be deduced by cosmic scholarly reasoning, for Hariot it could only be won through actual voyages of risk and adventure.

BACON AND THE RHETORIC OF EXPLORATION

The rhetoric of exploration and discovery, needless to say, was not limited to mathematical practitioners. The great voyages of the fifteenth and sixteenth centuries profoundly impressed early modern intellectuals, as it suggested to them that similarly unimaginable discoveries may yet be revealed in other fields of knowledge besides geography.[29] Francis Bacon was perhaps the most celebrated of the natural philosophers who sought to model their intellectual quest on the adventures of the great explorers. In *The Great Instauration* and the *New Organon*, he likened the search for knowledge in the natural world to a voyage of exploration in distant lands:

> The universe to the eye of the human understanding is framed like a labyrinth, presenting as it does on every side so many ambiguities of way, such deceitful resemblances of objects and signs, natures so irregular in their lines, so knotted and entangled. And then the way is still to be made by the uncertain light of the sense, sometimes shining out, sometimes clouded over, through the woods of experience and particulars.[30]

The search for truth, according to Bacon, must be conducted through dark and knotty entangled woods, aided only occasionally by glimpses of light that break through the cloudy skies.

Since the quest for knowledge is a voyage, the great maritime expeditions of the day are a natural model for Bacon's reform program:

> Nor must it go for nothing that by the distant voyages and travels which have become frequent in our times many things in nature have been laid open and discovered which may let in new light upon philosophy. And surely it would be disgraceful if, while the regions of the material globe—that is, of the earth, of the sea, of the stars—have been in our times laid widely open and revealed, the intellectual globe should remain shut out within the narrow limits of old discoveries.[31]

True natural knowledge, for Bacon, is an undiscovered country, "shut out" and concealed in an age when geographical knowledge is being expanded immeasurably. He concludes, "And therefore it is fit that I publish and set forth those conjectures of mine which make hope in this matter reasonable, just as Columbus did, before that wonderful voyage of his across the Atlantic."[32] Bacon promises to lead the way through uncharted waters, dark woods, and a labyrinth of knotted and entangled lines, to the prosperous land beyond. He is, in other words, the Columbus of the "intellectual globe."[33]

Bacon's actual involvement with the voyages of exploration was, in fact, rather limited. James I's policies were far more conciliatory toward Spain than those of his illustrious predecessor, and during his reign the Crown relinquished much of its patronage of the imperialist project. As Lord Chancellor, Bacon could do little more than voice his support for the various colonizing ventures taken up by the London merchant companies.[34] It is not perhaps surprising, therefore, if Bacon's writings show him to be less than an expert on maritime issues. His relative ignorance of actual maritime exploration is evident, for example, in his account of technological improvements in navigation. According to Bacon, whereas the sailors of the past sailing "along the shores of the old continent" could navigate solely by the stars, the development of the compass was necessary for the transoceanic voyages of the age of exploration.[35] In fact, Bacon here has the order of development reversed. The compass was invented around the twelfth century and was certainly in extensive use by the thirteenth, hundreds of years before Columbus ever set sail. Its usefulness was, in fact, greatly diminished in the great transoceanic voyages that Bacon had in mind, due to the effects of compass variation. The importance of celestial observations, contrary to Bacon's dismissal, grew immensely during this period.[36]

Bacon's elegant prose is undoubtedly far superior to Hariot's unpolished verses of the "Three Sea Marriadges." Nevertheless, Hariot's crude poem demonstrates an intimate and accurate knowledge of navigational techniques that Bacon completely lacks. The marriage of the "sun and star" refers to the difficulty of arriving at consistent navigational readings while basing the calculations on different celestial objects during the day and night. A similar problem arises from using different instruments, the "staff and sea astrolabe," to determine the ship's position. These were indeed serious and challenging tasks for the period's navigators. Most importantly, Hariot's reference to "charde and compass," refers to the challenge of producing a chart that will preserve true compass directions. It was precisely this problem that ultimately led to the development of the Mercator projection.[37] Significantly, Thomas Hariot worked and wrote on all these navigational problems.[38] In comparison, Bacon seems to be distinctly uninformed.

Bacon appears quite as ignorant of the history of the voyages as he is about navigational technology. Columbus, he writes "gave the reasons for his conviction that new lands and continents might be discovered besides those which were known before; which reasons, though rejected at first, were afterwards made good by experience, and were the causes and beginnings of great events."[39] Columbus, of course, never suggested the existence of any unknown lands, with the possible exception of some islands in the Atlantic. From first to last he always insisted that he had found the western route to Asia. The claim that he had discovered "new lands and continents" was shocking and unwelcome news to him, which he was at pains to deny to his dying day.[40]

It is impossible to know whether Bacon was unaware of the inaccuracies of his historical and technical references, or whether he simply did not much care. Quite likely he simply preferred the mythical Columbus, the prophet of rationality and progress, to the rather ambiguous historical Columbus who was staunchly medieval in his outlook. Either way, it is clear that the technical and historical details of the actual voyages of exploration were not his main concern. He was neither a sailor himself, nor a member of the supporting cast of financiers, shipbuilders, instrument makers, and mathematical practitioners that were closely associated with the expeditions. For Bacon, the voyages existed primarily as an abstract story—an inspiring tale of exploration and discovery that became his model for the search for knowledge.

Let us examine the basic tenets of Bacon's story of discovery, as it appears in quoted passages from *The New Organon*: great and wondrous lands lie hidden beyond the ocean, unknown and undiscovered. They are protected by the uncharted seas, and he who wishes to reach them must find his way through darkness and penetrate a labyrinth of conflicting signs. Access to the lands appeared quite impossible until Columbus and other great explorers arrived on the scene. They found a way through the intricacies of the labyrinth and "widely opened and revealed" the wondrous secrets of the land. All these elements amount to a familiar tale. Bacon's conception of the voyages was clearly shaped by the standard narrative of exploration, so prominent in the rhetoric of the English voyages. Essentially, Bacon here transfers the entire tale of discovery from its origin in English imperialist propaganda into the realm of natural knowledge.

The fact that Bacon utilizes the master narrative of exploration in his program for the reform of knowledge is clear evidence of the prevalence and success of imperialist rhetoric. Bacon was confident that his readers were familiar with the tale of exploration and would draw the correct conclusions from the analogy between natural knowledge and geographical discoveries. Great stores of knowledge await the enterprising discoverer, he suggests, who will find the way through the labyrinth of sense perceptions and ambiguous signs to the clear light of truth. The legendary success of the geographical voyages serves as a great promise that a way can indeed be found to the hidden mysteries of natural knowledge as well.

THE LABYRINTH OF MATHEMATICS

If Bacon's use of the discourse of discovery is evidence of its wide dissemination by the early 1600s, then it is hardly surprising to find the same rhetoric much closer to its imperialist source—among mathematicians. William Oughtred (1575–1660) was Bacon's younger contemporary, and like Bacon he was not intimately involved in the voyages of exploration. Early in his career Oughtred was apparently occupied with technical problems of navigation and even invented a new type of astrolabe.[41] By 1631, however, when he published the first edition of his best-known work, the *Clavis Mathematicae*, Oughtred had developed a disdainful attitude toward the work of mathematical practitioners and instrument makers.[42]

Oughtred's limited involvement in the technical aspects of exploration is probably no coincidence. The decline in royal patronage of exploration under the early Stuarts, along with the gradual institutionalization of mathematical studies at the universities, weakened the interdependency between mathematics and exploration.[43] But whatever the reason, it is clear that Oughtred's notion of the voyages was much like Bacon's. For both, the exploration of new lands was not a lived reality but rather an abstract story. Gone was the interest in the practical details of voyaging, which so occupied Dee, Hariot, and Wright; it was replaced by an imperialist tale of exploration and discovery.

But although the kind of use Bacon and Oughtred made of the imagery of exploration was rather similar, the subject matter to which they applied it was very different. For Bacon the expeditions served as a metaphor for the search for natural knowledge in general, through the use of a systematic empirical methodology. Mathematical knowledge did not figure prominently in his program, which stressed careful observation over abstract reasoning. Oughtred, in contrast, reserved his comments for mathematics alone. The true voyage of discovery for him was the search for mathematical truth.

In the preface to the English edition of the *Clavis*, Oughtred described his experience when he first studied the classical mathematicians "Euclides, Archimedes, Apollonius Pergaeus that great geometer, Diophantus, Ptolemaeus and the rest."[44]

> Truly when I was conversant in reading their bookes, and with wonder
> observed their most witty demonstrations, so skillfully framed out of principles,
> as one would little expect or thinke, but laid together with divine artifice: I

was even amazed whence possibly any power of imagination should be able to sustaine so immense a pile of consequences, and cause that so many things, so far asunder distant, could be at once present to the minde, as with one consent joyne and lay themselves together for the structure of one argument.[45]

The classical mathematical texts appeared to Oughtred as a perfect web of relationships, in which each element was interconnected with all other elements in innumerable surprising ways. He then proceeds to explain his method of dealing with this intractable web: "Wherefore that I might more clearly behold the things themselves, I uncasing the propositions and demonstrations out of their covert words, designed them in notes and species appearing to the very eye."[46] Oughtred here proposes to extract the "covert" elements from the web of words and examine them closely with his "very eye." The reference to "notes and species" here is not coincidental: it points to the most significant innovation of Oughtred's work, namely, his introduction of new algebraic symbols and his insistence on their superiority over textual descriptions of mathematical operations. For Oughtred, "uncasing the propositions and demonstrations" involved, first of all, restating them in mathematical symbols. The confusing web of words, which comprises the classical texts, is wondrous in itself, according to Oughtred. Unfortunately, it obscures the real nature of the "things themselves." In order to truly understand the mathematical classics, one must penetrate through the opaque words and logical connections and reach the covert truth beyond them.

As was the case with Bacon, the labyrinth provides Oughtred with an eloquent metaphor of his purpose in writing the *Clavis*. The purpose of his "Key," he writes, "is to reach out to the ingenious lovers of these Sciences, as it were *Ariadne*'s thread, to guide them through the intricate Labyrinth of these studies."[47] Elsewhere he denigrates the work of practitioners who do not search deep enough in the mathematical maze. What they call "practice," he contends, "is in reality mere juggler's tricks with instruments, the surface so to speak, pursued with a disregard of the great art."[48]

The Cretan labyrinth, which for Bacon represented the search for natural knowledge in general, is for Oughtred a metaphor for the study of mathematics. It is vast and confusing, presenting a smooth impenetrable "surface" to the aspiring student of the subject. Some practitioners are evidently content to remain on the surface level, satisfied with mere superficial "juggling." Not so Oughtred: he casts himself in the role of a pathfinder, who locates a passage through the intractable labyrinth and leads the way to true mathematical knowledge.

The vision of knowledge adopted by Oughtred is highly reminiscent of

Bacon's views. Mathematical knowledge, for Oughtred, is a hidden treasure, concealed beyond the obscure texts of antiquity and the classical synthetic form of presentation. The enterprising scholar must break through this obstructive screen and find his way to the pure knowledge of mathematics. All the elements of the master narrative of exploration are present in this vision: the promised land beyond, represented here by pure mathematical truth; the labyrinthine obstacles on the way; and, finally, the enterprising explorer, Oughtred himself, who finds a way through all difficulties and leads his companions to the elusive treasure.

As if to leave no doubt as to his source of inspiration, Oughtred's correspondence is riddled with references to the study of mathematics as an enterprise of exploration and empire. Being an enthusiastic advocate of Cavalieri's indivisibles, he wrote to Robert Keylway that he is confident that "a great enlargement of the bounds of the mathematical empire will ensue" from Cavalieri's method. Unfortunately, he told Keylway, he himself was too old and too demoralized by the civil war raging around him to embark on the voyage himself: "Being more stept in years, daunted and broken by the sufferings of these disastrous times, I must content myself to stay home and not put out to any foreign discoveries."[49]

Significantly, the same imagery can be seen at work not only in Oughtred's own letters but also in those he received. William Robinson wrote Oughtred in 1636 that "The abyss of these sciences is inexhaustable . . . and when all is done, there will be a terra incognita for mathematicians of after ages to sail into."[50] The mathematician as a resourceful explorer of hostile terrain is clearly the guiding metaphor of the study of mathematics for Robinson as well.

Oughtred's rhetoric seems to represent a late stage in the association of mathematics with the imperialist project. Oughtred himself was only marginally involved in the voyages. He utilized the abstract discourse of the explorers, but he was not much interested in the actual technical aspects of oceanic navigation or cartography. It is probably no coincidence that we find a rather similar situation at the other end of the historical spectrum, at the very beginning of the association of mathematical practitioners with English exploration. Robert Recorde was likely the first mathematician to be involved in the voyages. Like Oughtred many years later, his ties to the exploration project can be characterized as broadly ideological rather than technical.

Unlike his successors, Dee, Wright, and Hariot, Robert Recorde did not serve as an advisor to any actual voyages. This is clearly evident in the series of textbooks he published in the 1540s and 1550s. His most significant contribution to the literature of navigation came in 1556, when he published a book entitled *The Castle of Knowledge* on the construction and use of the sphere. The book, according to E. G. R. Taylor, was written specifically to further the search for Cathay and clearly demonstrates Recorde's interest in the cosmology of his day.[51] It was, furthermore, one of the few volumes known to have been in the library of Frobisher's northwest expeditions twenty years later. Despite its maritime topic, however, the *Castle* was no navigational manual. It was a learned and scholarly work, based almost entirely on classical and medieval sources, complete with a critique of the schoolmen's knowledge of the Greek language. This was hardly the kind of work that Tudor mariners would find useful in navigating the high seas.[52] Recorde wrote another treatise on navigational matters entitled *The Gate of Knowledge*, about the use of the quadrant and other mathematical instruments. This work, however, is lost and was probably never published.[53]

What is true of the *Castle* is also generally true of *The Pathway to Knowledge*, Recorde's 1551 book on geometry. Here, as elsewhere, Recorde expressed a keen interest in exploration, but was sorely lacking in practical details. The preface contains a poem by Recorde, in which he lists maritime travel as the first of many uses of geometry:

> Sith Merchauntes by shippes great riches do winne,
>   I may with good righte at their feate beginne.
> The Shippes on the sea with Saile and with Ore,
>   were firste founde, and stylle made by Géométries lore,
> Their Compas, their Carde, their Pullies, their Ankers,
>   were founde by the skill of witty Geometers.[54]

The poem registered Recorde's enthusiasm for the maritime ventures of his day. He clearly believed that mathematics had an important role to play in the voyages, and suggested that great riches could be won through its proper use. When it comes to the practical applications of geometry, however, the *Pathway* falls short and does not discuss the actual ways of applying geometry to shipbuilding or navigation.

An exception to this rule emerges in Recorde's dedication of his 1557 book on algebra, *The Whetstone of Witte*, to the Muscovy Company.[55] Recorde was apparently aware of the practical shortcomings of his work, because his preface contains a concrete promise: "I will for your pleasure, to your coumforte, and for your commoditie," he wrote, "shortly set forthe soche a booke of navigation, as I dare saie, shall partly satisfie and contente, not onely your expectation, but also the desire of a greate number beside."[56] Such a book, unfortunately, has not survived and quite likely was never written. On the whole, despite Recorde's stated commitment to geographical exploration and his serious interest in mathematical navigation, his association with the voyages remained ideological and rhetorical rather than technical.

Much like Oughtred years later, Recorde was known more for his teaching skills than for his mathematical innovations. His elementary textbook of arithmetic, *The Grounde of Artes*, was first published in 1543 and proved to be the most popular "Arithmetic" in English for more than a century. Its last edition appeared as late as 1699 and was in continual use well into the eighteenth century.[57] Although the *Grounde* was the only one of Recorde's books to achieve such popularity, while his other writings were generally forgotten, for its author it was only the first step in a larger endeavor. He describes his vision in an introductory poem to *The Castle of Knowledge*:

AN ADMONITION FOR THE
orderly trade of studye in the Authors Woorkes,
appertaiyning
to the mathematicalles

The grounde is thought that steddye staye
Where no foote faileth that well was pyghte:
Whereon who walketh by certaine waye,
His Payse is lyke to prosper ryghte.

1. The *Grounde of Artes* who hathe well tredd,
   And noted well the slyppery slabbes,
   That may him force to slyde or falle,
   He hathe a staffe to staye withall.

2. Then if he trade that *Pathwaye* pure
   That unto knowledge leadeth sure:
   He maye be bolde tapproche *The Gate*

3. *Of Knowledge* and passe in thereat.
   Where if with *Measure* he doo well treate:

4. To *Knowledges Castle* he maye soone get.
   There if he travaile and quainte him well.

5. The *Treasure of Knowledge* is his eche deale.

5. This *Treasure* though, that some wold have,

3. Whiche *Measures* friendshippe do not crave,

2. Nor walke the *Patthe* that leadeth the waye,

1. Nor in *Artes grounde* have made their staye,
   Thoughe bragge they maye, and get false fame,

4. In *Knowledges Courte* thei never came.[58]

Recorde's poem enumerates his various works, and pinpoints the place of each one in his grand project. The highest form of mathematics is the *Treasure of Knowledge*, a book that was never published and probably never completed. Modern scholars have speculated that it was either the promised work on navigation, or an advanced cosmography following up on the *Castle*.[59] There is no way of knowing for sure. The significant aspect for us is that Recorde viewed it as the culmination of mathematics and considered it a guarded treasure. In order to reach the *Treasure* one must enter the *Castle*—Recorde's work on the sphere. The entrance to the castle is guarded by a *Gate*, Recorde's treatise on instruments, which can be approached by a *Pathway*—the title of his book on geometry. The whole edifice stands on the *Grounde of Artes*, that is, elementary arithmetic, the starting point for anyone seeking mathematical knowledge.

The central elements of this story should be familiar by now, as they share certain crucial aspects of the standard tale of exploration. The hidden riches protected by formidable defenses are familiar themes from the travel narratives of Frobisher, Hariot, and others. So is the narrow pathway, which penetrates through the defenses and provides access to the treasure in the interior. Most importantly, the enterprising explorer of our familiar tale appears here in the guise of the student of mathematics, who will not be deterred by the great difficulty of the subject matter. All of these elements seem to combine into a story that is very much like the familiar narrative of exploration.

And yet it must be acknowledged that despite the similarities, Recorde's vision of mathematical knowledge as a guarded treasure to be possessed does not quite qualify as a tale of exploration. Challenging an enterprising hero to penetrate the castle and acquire great riches is not enough to characterize a story as a narrative of discovery. What is missing is a sense of novelty, the realization that the treasures thus acquired are completely new, previously unknown and undiscovered. The mathematical riches that are to be had in Recorde's "Castle of Knowledge" are not new discoveries, previously unsuspected mathematical gems. They are, rather, traditional mathematical methods and results, which had been, for the most part, known since antiquity. Recorde, after all, was writing textbooks, designed to instruct the uninitiated in the well-established mathematical arts. His purpose was not to uncover new truths, but rather to familiarize a wider public with known ones. The hero of the assault on the mathematical castle is not a learned "geometer" in search of knew knowledge, but a mathematical novice, seeking to overcome his own ignorance.

In characterizing Recorde's view of mathematics as expressed in this poem, one must go back to the original sources of the standard tale of exploration.

This narrative, it will be recalled, combined elements from the crusading tradition with aspects of the chivalric romances. Recorde's vision seems to contain much of the crusading spirit, but rather little of the imagery of the knight-errant. Like a crusader besieging Jerusalem, Recorde's mathematical hero sought to conquer the fortified citadel and partake in the pure and unsullied knowledge contained therein. He did not seek new adventures at the far reaches of the earth like the knight-errant of romance, but rather the true and original source of all knowledge, in the manner of a crusader.

Elsewhere in Recorde's body of work, the search for new and unsuspected knowledge takes on a greater prominence. In the preface "To the Gentle Reader" of *The Pathway to Knowledge* Recorde apologizes for any imperfections that may be contained in his work: "Excuse me, Gentle Reader if oughte be amisse, straung paths are not troden al truly at the first: the way must needs be comberous, wher none hathe gone before. . . . I will not cease from travaile the pathe so to trade, that finer wittes maie fashion themselves with such glimsinge dull light, a more complete woorke."[60]

The significance of this passage becomes clear when it is compared to Recorde's address to the Muscovy Company in *The Whetstone of Witte*: "If you continue with corage, as you have well begon, you shalle not onely winne greate riches to your selves, and bryng wonderfull commodities to your countrie. But you shall purchase therewith immortall fame, and be praised for ever, as reason would: for opening that passage, that shall profite so many."[61] The parallels between the two passages are unmistakable. Recorde views his own mission and that of the explorers of the Muscovy Company in similar terms. Both are to enter courageously into unknown territories and find their way through strange and unfamiliar lands. They will bring light to darkness and will be rewarded with "treasure" and "great riches." Even more importantly, Recorde's mathematical work, as well as the Muscovy Company's voyages, will open a new way, a passage for others to follow. Recorde here fashioned himself as a trailblazer, leading others through the untrodden paths of mathematics, much as the sailors of the Muscovy company opened new paths of travel and commerce.

Recorde's vision of mathematical knowledge and the ways to attain it is here similar to that of Oughtred decades later. In both their accounts, mathematical knowledge lies hidden and protected by numerous barriers. For Oughtred, the chief obstacles were the classical synthetic form of presentation, as well as the absence of algebraic symbols in the ancient texts. This, according to Oughtred, obscured the actual form of mathematical reasoning followed by the ancients. For Recorde, the main difficulty, it seems, was the inherent complexity of the mathematical subject, which must therefore be approached gradually and systematically. For both, the mathematician's role was to penetrate through great obstacles and find a path to the treasures of mathematical knowledge. In the end, it seems, even Recorde's view of mathematics came remarkably close to the familiar imagery of exploration and discovery.

Thomas Blundeville (fl.1560–1602) represents a very different stage in the association of mathematics with the voyages than either Recorde, his predecessor, or Oughtred, a generation later.[62] For Blundeville, maritime exploration was not an abstract tale of discovery, as it was for Oughtred. Nor was it a desirable goal, constantly aspired to but never attained, as it seems to have been for Recorde. When Blundeville published his popular *Exercises* in 1594, the Elizabethan exploration enterprise was at its height.[63] Drake's circumnavigation (1576–80) was a relatively recent memory. Walter Raleigh's colonizing effort in Virginia had ended in apparent disaster, but he was not deterred: in 1594 he sent out his first scouts to begin the search for the lost kingdom of El Dorado. During the entire period English seamen regularly preyed on Spanish shipping and raided the settlements of the Spanish Main. Under these circumstances, navigation and cartography became an immediate practical concern for an increasing number of oceangoing sailors and navigators, as well as a central professional concern for a growing class of mathematical practitioners.

Blundeville and his mathematical contemporaries viewed mariners not as mythical heroic figures, but rather as the immediate target audience at whom their treatises were directed. It could well be argued that during Blundeville's generation, which also included Thomas Hood, Thomas Hariot, and Edward Wright, among others, the association of mathematicians with the imperialist project reached its peak.

For Blundeville, the main purpose of mathematical studies was clear: it was to provide the necessary technical assistance to transoceanic voyages. This position is evident in his famous *Exercises*, which included a discussion of elementary arithmetic, as well as treatises on globes, maps, and the principles of navigation. It also contained the first publication of Wright's tables of meridional parts, necessary for the construction of Mercator projection maps.[64] The close connection that Blundeville draws between mathematics and seafaring was not, however, limited to the technical content of the book, but was plainly stated in its title page. The full title of this work runs as follows:

> M. Blundeville, His Exercises, containing sixe Treatises, the titles whereof are set downe in the next printed page: which Treatises are verie necessarie to be read and learned of all yoong Gentlemen that have not beene exercised in such disciplines: and yet are desirous to have knowledge as well in Cosmographie,

Astronomie, and Geographie, as also in the Arte of Navigation, in which Arte
it is impossible to profite without the helpe of these, or such like instructions.
To the furtherance of which Arte of Navigation, the said M. Blundeville specially wrote the said Treatises, and of meere goodwill doth Dedicate the same
to all young Gentlemen of this Realme.[65]

The same theme is repeated in several places in the introductory sections
of the work. "I greatly rejoyce," Blundeville writes in the opening of his preface to the reader, "to see so many of our English Gentlemen . . . so earnestly
given to travel as well by sea as land, into strange and unknowne countries;
& specially into the East & West Indies, following therein the good example
of divers worthy knights & Gentlemen, that have ventured their lives to discover strange countries to the great honour of their countrie, and to their
owne immortall fame." Blundeville here describes the imperialist enterprise
in terms of its standard rhetoric: courageous and enterprising men set forth
into undiscovered countries, winning fame and fortune for themselves and for
England. He then defines the purpose of his own work: "And because that
to travell by sea requireth skill in the Art of Navigation, in which it is impossible for any man to be perfect unlesse he first have his Arithmeticke, and also
some knowledge in the principles of Cosmographie . . . I thought good
therefore to write the Treatise before mentioned, to serve as an introduction
for such young Gentlemen."[66] Later in the introduction, Blundeville explains
that he wrote the treatise for "those that are desirous to studie . . . speciallie
the Arte of Navigation." He concludes by stating that the work would be useful for "specially such seamen as have some tast of Arithmeticke, without
which no good almost is to be done in any science."[67]

Blundeville was so committed to the natural connection between mathematical studies and navigation, that even his book on Ptolemaic cosmology,
The Theoriques of the Seven Planets, contained the following exhortation in
its title: "A booke most necessarie for all Gentlemen that are desirous to be
skilfull in Astronomie, and for all pilots and Sea-men, or any other that love
to serve the Prince of the Sea, or by the Sea to travell into forraine
Countries."[68] The usefulness of the detailed discussion of Ptolemaic epicycles
and eccentrics, contained in this tract, for purposes of navigation is doubtful, to say the least. The significant point is that Blundeville considered any
discussion of mathematics to be fundamentally aimed at assisting seafaring.
For Blundeville, seafaring and navigation were the primary goals of any
mathematical studies.

Others among the community of London mathematical practitioners
agreed. John Aspley, who in 1624 published a tract entitled *Speculum
Nauticum* on the nautical uses of a "plain scale" he invented, clearly shared

Blundeville's conviction that the primary purpose of mathematics is to aid in oceanic voyages. The *Speculum* was dedicated to Trinity House in Debtford Strand, an institution that served as a semiofficial headquarters and school for navigators. He prefaced the work by promoting the study of mathematics as a patriotic duty: English sailors who "are unto this Iland as a wooden wall, the Sea chariots, and the horses of England" deserve to enjoy the benefits of mathematical study. "These, I say, may claime justly to the fruits of our labours . . . which have not altogether been abhorrent from the mathematical studies . . . to further that so much deserving science of Navigation."[69] For Aspley, as for Blundeville, seafaring and navigation were the primary purpose of mathematical studies. In London, mathematics and the voyages went hand in hand.

EDWARD WRIGHT AND THE KEY TO MATHEMATICS

Edward Wright (1561–1615) was a far more creative mathematician than either Blundeville or Aspley, but he clearly shared his contemporaries' views on the relationship of mathematical studies to oceanic voyages.[70] Wright personally took part in the Earl of Cumberland's raid on the Azores in 1589 and published a detailed account of the voyage.[71] On his return he composed a technical manuscript in which he suggested some major changes in navigational and cartographical practices. Wright was eventually persuaded to publish the treatise ten years later, after parts of the manuscript were printed by others (including Blundeville) and he became concerned that he would not receive due credit for his work. The resulting book was *Certaine Errors in Navigation*, Wright's most celebrated work, which firmly established his reputation as a leading mathematician and cartographer.[72]

While *Certaine Errors* was composed of several sections, dealing with topics such as compass variation and the declination of the sun, the largest and most important part dealt with the problem of cartographic projection. The marine charts most commonly in use at the time were known as "plane charts." Ignoring the curvature of the earth, such maps depicted both latitudes and meridians as equally spaced parallel lines. This, of course, was only a rough approximation, since whereas the lines of latitude are indeed parallel to each other everywhere, the meridians converge with the rising latitude until they meet at the poles. The plane chart was quite adequate as long as the mapped area was relatively small and located in the lower latitudes, where the meridians are approximately parallel. Once those limits were exceeded, as they regularly were in the discovery voyages of the period, the distortions of such charts grew alarmingly, making them quite useless.

The shortcomings of the plane chart became obvious when they were used as navigational aides in transoceanic voyages. If a navigator were to set a course in a given direction on a plane chart, he would be greatly deceived. His north-south departure would be measured accurately, since the latitudes are correctly depicted in his chart as equally spaced parallel lines. The east-west departure, however, would be grossly underestimated especially in the higher latitudes, since the meridians in fact converge, while on the chart they remain parallel and regularly spaced. The challenge faced by cartographers of the day was to produce a marine chart that would avoid these problems. On such a map, if a navigator wished to chart the course of a ship sailing in a fixed direction, he would merely need to draw a straight line at the given direction, starting from the point of departure. The cartographers' goal, in other words, was to produce a map that would preserve true directions.[73]

The first to produce such a map was Gerard Mercator in his world map of 1569. The basic principle of his projection was simple: rather than depict the meridians as converging, Mercator kept them as equally spaced parallel lines just as they were on the plane chart. In order to compensate for the distortions, he systematically increased the distance between the latitudes as they approached the poles. The ratio between degrees of latitude and longitude at any given point was thus preserved in this arrangement. A fixed course on such a map crosses the correct number of meridians for each degree of latitude traversed, and directions are therefore preserved.

While the basic idea of this projection was recognized by the mathematical practitioners of the time, Mercator never actually explained the method of calculating the appropriate distance between the parallels at each given latitude.[74] In 1599, when *Certaine Errors* came out, it was the first systematic exposition of the construction of a Mercator projection map. In *Certaine Errors*, Wright was the first to explain the basic principles for calculating the increasing distance between the parallels. He then used his method to calculate the "meridional parts" tables, necessary for the construction of a Mercator map.

A more detailed account of Wright's mathematical technique, known as "the addition of secants," is given in Chapter 5. More significant, for now, is the rhetoric and imagery that he employed in presenting his new method. No man, he argues, can deny the excellence of the science of navigation, or its profitableness. Navigation, Wright claims, made possible "the more wonderfull discoveries in this our age, made to the furthest parts of all the earth, and rounde about the whole compasse of the same, whereby we have been made partakers of the most rare and richest commodities and treasures of the utmost Indies, and Islandes of the world."[75] Navigation, then, opens the way to "commodities and treasures" hidden in "the furthest parts of all the earth."

Unfortunately, according to Wright, the mariners' charts have not kept pace with the new discoveries. Instead of assisting the navigators, they have become a hindrance: "The sea chart, the best meane the mariner hath to knowe the course from place to place . . . is so faulty in the very foundation and groundworke thereof (that is in the geometricall lineaments of the meridians, paralels, and rumbes described therein) that hereof there may arise so grosse error, as may cause the mariner to misse one, two, yea three whole points of the compasse."[76] The arrangement of the parallels and meridians on the plane chart, Wright claims, makes them worse than useless for the navigator. He then concludes, "It cannot otherwise be but that the ordinary charts are in many places much like an inextricable labyrinth of error, out of which it will be very hard for a man to unwinde himself."[77] The familiar metaphor of the labyrinth, popular with Bacon and Oughtred (as well as with Hariot) should serve as a clue that a narrative of exploration is unfolding in Wright's rhetoric. Already we have great hidden treasures in the interior, and the way hither turned into a labyrinth by the errors of the plane chart. Next, Wright provides a brief account of mathematicians' efforts to break through the formidable obstacles. To the difficulties of the mariners, he writes,

> We may adjoyne the experience of the best Hydrographers of our time: who
> dayly making their Charts after the accustomed manner with streight lined
> rumbes and degrees of latitude, everie where equall, have found such difficulties
> in labouring to bring their marine descriptions to some due correspondence
> of trueth in the courses, heights and distances, that tyred herewith in the end,
> they have holden it for impossible, to make the chart agree in all these with the
> globe.[78]

The "labyrinth of error" has evidently defeated the best efforts of the hydrographers of the age. They have despaired of penetrating it and concluded that the problem was insoluble.

Wright himself does not share the hydrographers' pessimism and comments, "Wherein notwithstanding they erre, by making too general a conclusion, in houlding that to be simply impossible, which cannot be done by such a way & meanes as they know and use."[79] The hydrographers' real problem, according to Wright, is that they simply do not know the "way" that leads to the solution of the problem. Wright himself, of course, claims to have found this passage. After explaining his technique in detail, he pauses for a final evaluation of his achievement: "Now then wee have an easie way layde open for the making of a table . . . whereby the meridians of the Mariners Chart may most easily and truly be divided into parts."[80] Wright encountered the labyrinth that had defeated his predecessors and found an "easie way" through it.

Wright here presents his achievement as clearly on a par with those of the geographical explorers. Like Frobisher in the Canadian arctic, or Raleigh in Guyana, he sets out in search of great riches. In both the geographical and the mathematical case, these treasures are protected by major obstacles. The geographical explorers must break through mountains, dark forests, or ice barriers; the mathematical adventurers must penetrate "a labyrinth of error" that seems insoluble even to the best practitioners. Wright, like his illustrious voyaging contemporaries, claims to have broken through all obstacles and found the open way to the hidden treasures. He is, in other words, a traveler in his own right, exploring the uncharted territories of mathematics and searching for a "northwest passage" to the secrets hidden within.

The image of Edward Wright as an explorer in the lands of mathematics reappears in the other work for which he is famous—his English translation of Napier's book on logarithms. John Napier's Latin *Mirifici Logarithmorum Canonis Descriptio* was published in 1614 in Edinburgh, and in stark contrast to London mathematical publications, it contained no reference to the utility of the new technique for seafaring.[81] Wright's translation appeared only two years later in London and bears all the markings of the English maritime mathematical environment. *A Description of the Admirable Table of Logarithmes* was dedicated "To the Right Honourable and Worshipfull Company of Merchants of London trading to the East Indies."[82] Since Wright himself died in 1615, it was left to his son, Samuel, to praise the East India Company for its "continuall imployment of so many Mariners in so many goodly and costly ships, in long and dangerous voyages, for whose use (though many other wayes profitable) this little booke is chiefly behoovefull."[83]

Napier, the Scottish country laird, never considered navigation to be an important concern, and certainly not the primary aim of his work. For Samuel Wright, with his close ties to the English imperialist enterprise, the connection of logarithms to the voyages seemed self-evident. Important mathematical innovations, it seemed to him, would naturally find their primary use in the hands of mariners.

As before, Wright presents his mathematical work as not only extremely useful for the mariners but also as a voyage of discovery in its own right. This theme took on concrete form in an introductory poem to Wright's *Description*, composed by John Davies of Hereford. The secret of logarithms, according to Davies, was "enwomb'd with clouds of mystery" until, thanks to Napier and Wright, they were "produc'd to light." Their achievement was that they

> ... for *Mathematicks* found this *Key*,
> To ope the lockes of all their *Misteries*,
> that from all eyes so long *concealed lay*.[84]

The theme of finding a key to the secrets of mathematics is already familiar from Oughtred's *Clavis Mathematicae*. The precise nature of the mathematical secrets is, of course, different in each case, but the discovery narrative is the same. The mathematical explorer breaks through "clouds of mystery" to bring to light truths that have previously been hidden.

Davies was not satisfied at making generic use of the rhetoric of geographical exploration. He proceeded to describe Wright literally as a skilled navigator sailing the rough waters of mathematical knowledge.

> Wright (ship-wright? no; ship-right, or righter then,
> when wrong she goes) lo thus, with ease, will make
> Thy Rules to make the ship run rightly, when
> She thwarts the *Maine for Praise or profits* sake.[85]

The image of Edward Wright in Davies' verses follows closely the pattern Wright himself established in *Certaine Errors*. In both places Wright is an enterprising explorer, seeking to unveil the hidden treasures of mathematical knowledge. Wright fashions himself (and is fashioned by others) in the role of the "discoverer" in the standard narrative of exploration.[86]

John Tapp (fl.1596–1631), in his treatise *The Pathway to Knowledge* of 1613, neatly synthesizes the various relations drawn by mathematicians between their own field and voyages of exploration.[87] Tapp was a well-respected teacher of mathematics and navigation, who gained his reputation by issuing an edition of Richard Eden's translation of Martín Cortés's classic *Arte of Navigation* in 1596. He followed this by issuing an immensely successful nautical almanac in 1602 entitled *The Seamans Kalendar*.[88] *The Pathway to Knowledge* was a book of elementary arithmetic, combined with certain fundamentals of algebra.

The introductory sections of Tapp's book represent both aspects of the connection made between mathematics and oceanic navigation by the mathematical practitioners of the time. On the one hand, like Blundeville, Aspley, and Wright, Tapp emphasizes that the study of mathematics should be primarily oriented toward nautical improvements; on the other, like Oughtred, Recorde, and Wright (again), he uses the tale of exploration as metaphor for mathematical studies themselves. In Tapp's work, as in Wright's, the two strands combine to promote mathematics as inseparable from the exploration enterprise of the time.

Tapp follows the precedent set by his predecessors, but tops them all by dedicating his treatise to four different voyaging companies: the East India Company, the Muscovy Company, the Northwest Passage Company, and the Virginia Company. This dedication in itself strongly suggests that the intended audience for Tapp's arithmetic is the oceangoing voyagers of his day. Thanking his patron Sir Thomas Smith for his support, Tapp proceeds to praise him for his sponsorship of "Arts (but most especially *Navigation*) which is the most noble stem and profitable branch that ever sprang from them." Smith's employment "of shippes, Mariners, and others, not only for setled trades but also for new discoveries, [is] to the great benefit of many thousands . . . and a lasting glory to our nation."[89] For Tapp, navigation and discovery of new lands are the most glorious endeavors imaginable.

This grandiose rhetoric is then followed by a complaint: mathematical lectures are being offered for the benefit of mariners in London, Tapp claims, but few seem to be taking advantage of them.[90] "What good," he asks, "doth these publique readings which hath now beene a reasonable time continued in this Cittie, with greate charge to good purpose, but little profit as may be guessed, by the little audience which doe commonly frequent them." The problem, he suggests, lies in the technical mathematical nature of the readings: "And for the Arts there taught, I meane the mathematiques, the practisers thereof are few, in respect of those that are practisers and professors of *Navigation*."[91]

The solution, according to Tapp, is to tailor the lectures more precisely to the technical needs of navigators. "But were there a lecture of *Navigation*, a profession which a multitude of people make their onely living by . . . there is no question to be made of a very sufficient Auditory and great benefit to be reaped thereby."[92] He then concludes by stating confidently that new publications and public readings in mathematics have already produced more capable mariners "within these few yeares, than in former times can be found almost in the whole age of a mans life."[93]

According to Tapp, the true purpose of mathematics, then, is to provide technical assistance to mariners and navigators. Lectures that focus too much on pure mathematics will be of no practical use to them, will lose their audience, and will become a waste of time and resources. Even the present faulty lectures, he claims, in as much as they touched on practical navigational matters, have had remarkable educational results. How much more would this be the case if the mathematical lectures were designed to fulfill their proper and legitimate aim—supplying practical technical assistance to oceangoing seamen.

If the voyages were the main purpose of the study of mathematics, then they were also a metaphor for the pursuit of mathematics itself. The very title of the work, *The Pathway to Knowledge*, seems to suggest this. The title was, of course, used previously by Recorde, for whom the "pathway" was an intermediary stage between the "ground" and the "gate of knowledge," which opens the way to the "castle" and the "treasure."[94] While Tapp is not as elaborate as Recorde in his depiction of the progress toward the hidden secrets, the title does suggest that knowledge lies "beyond" and can only be approached by a proper "path."

Tapp then follows this by more explicit references to the study of mathematics as a voyage of exploration. Addressing Sir Thomas Smith and thanking him for his patronage, Tapp writes, "[Arithmetic] being the primary or first path-gate, or entrance, into the other Mathematiques, as also a principal assistant to *Navigation*, which is the chief and most effectuall branch produced from the naturall sap and nourishment of the former, where should it find a safer harbor to ride in, against the tempestuous stormes of turbulent depravors."[95] The underlying tale is rather similar to the one used by Oughtred in describing his attempts to decipher the writings of the ancients, or Wright in penetrating the "labyrinth of error" of the plane chart. The secrets of mathematics lie hidden beyond a "gate," and Tapp points to the open passage leading to them, that is, the "pathway" of arithmetic. The voyage to knowledge, however, is a difficult one, and Tapp must take advantage of the "safer harbour against tempestuous stormes" offered to him by his patron, Sir Thomas Smith.

The nature of the obstacles varies greatly from Oughtred to Wright to Tapp. The latter seems to complain about certain detractors of his work, which would prevent him from receiving his due recognition. It is not clear whether Tapp is referring to any particular critics (which would seem unlikely for a work on arithmetic), or whether he is simply using a standard rhetorical ploy in attempting to secure Smith's patronage. The significant point is the narrative that Tapp utilizes in describing his endeavor as a hazardous voyage toward knowledge. Tapp, like Recorde and Wright before him, and Oughtred later on, uses the standard tale of exploration and discovery to characterize his own mathematical work.

Recorde, Blundeville, Wright, Aspley, Tapp, and Oughtred provide but a sample of the mathematicians active in the early years of English geographical exploration.[96] They do, however, establish a pattern of discourse that existed among them and their peers. The practitioners associated mathematics closely, and sometimes exclusively, with navigation and the imperialist enterprise. They adopted its language, rhetoric, and narrative and then applied it to their own enterprise—the search for mathematical knowledge. The study of mathematics became, in their hands, a voyage of discovery in its own right, on a par with the naval expeditions of their patrons.

Elizabethan mathematicians undoubtedly possessed other ways of describing their work as well. John Dee possessed a vision of mathematics that was very different from the "search and discover" version elaborated here, to give but one example. It is clear, nevertheless, that "Exploration Mathematics" was common, prevalent, and dominant among the mathematical practitioners of the time. When in the following chapters we find Hariot referring to a "labyrinth" and to the passage to the hidden secrets of nature, his references will be clear. In utilizing those images, Hariot was taking part in the common discourse of his professional peers and applying the standard narrative of exploration to his mathematical work.

CHAPTER 4

Seeing the Truth: Hariot and the Structure of Matter

SEEING AND KNOWING

During his lifetime, Thomas Hariot was reputed to hold strong and provocative philosophical views. Persistent rumors described Hariot as a skeptic and a materialist, a leading member of Sir Walter Raleigh's School of Atheism.[1] It is therefore somewhat surprising to note that in all the manuscript papers Hariot left behind, traditional philosophical discussions and comments are in short supply. Hariot's papers are filled with notes on an extremely wide variety of issues: complex mathematical calculations are found side by side with discussions of the proper division of watches on an oceangoing ship and an estimate of the number of generations elapsed since creation. But in this vast array of topics, abstract philosophical speculation is almost completely absent.[2]

Only in one place does Hariot clearly articulate a stance that can be considered purely philosophical. On the reverse side of a page concerned with the laws of free-falling bodies, Hariot jotted down, "The truth when it is seen is knowne without other evidence."[3] From our modern point of view, this simple pronouncement could be taken to mean nothing more than a scientist's proclamation of his faith in his own intuitions. According to the modern scientific ethos, creative minds are often distinguished by their intuitive grasp of the truth, which circumvents the normal channels of reasoning. The laws of hydrostatics, according to legend, dawned on Archimedes while he was lounging in the public baths of Syracuse, and Newton reputedly grasped the principle of universal gravitation after watching the proverbial apple drop to the ground. No scientist would dispute that a great deal of hard work is later required to fill in the gaps of reasoning left by the great intuitive leap. By then, however, the essential truth of the proposition is already assumed.

This vision of scientific discovery has been adopted by certain strains of twentieth-century philosophy of science, which gave it a more formal guise as the distinction between the "context of discovery" and "the context of justification."[4] Only the latter is governed by strict logical considerations, while the former is left to the genius and imagination of the individual scientist. Karl Popper, Hans Reichenbach, and many a modern scientist would agree with Hariot's proposition: for true scientific visionaries, the truth is indeed known once it is seen, before other evidence can be marshaled.

Yet however mundane Hariot's proclamation would seem to us, to a contemporary of Hariot it carried profound and far-reaching implications. In early modern Europe, private intuition seemed an extremely precarious basis for the establishment of truth. Traditionally, true knowledge was based on the interpretation of a canon of ancient texts, which included scripture and the corpus of the ancient philosophers and church fathers. The underlying assumption was that all relevant knowledge was already in existence and was contained within the prescribed canon. The search for truth, therefore, consisted of the proper application of the wisdom contained within the bounds of these volumes to the problem at hand. If, as was often the case, the canonical texts were in conflict with each other, the difficulties would be resolved through the scholastic practice of disputation. Truth, in other words, was arrived at not through new discoveries but through hermeneutics—the detailed interpretation of authoritative texts.

When viewed against this epistemological background, Hariot's proclamation seems far less innocent than it appears at first glance to the modern reader. The proper way to knowledge, Hariot claims, is not through reading and interpreting, but through the direct, unmediated act of "seeing." Personal observation, he suggests, can completely circumvent the need for any further evidence. The eyewitness—he who sees—is, for Hariot, the proper bearer of true knowledge, surpassing any alternative sources of authority. Hariot's deceptively simple statement is, in fact, a direct challenge to the traditional order of knowledge.[5]

Hariot was not alone, of course, in contesting the traditional system of authority. The controversy between those who adhered to the established cannon and those who promoted new discoveries, often dubbed the "ancients" and the "moderns," persisted in European intellectual life throughout the sixteenth and seventeenth centuries. The dispute pitted the traditionalists, who maintained that seemingly new discoveries were already known to the ancients and contained in their texts, against the moderns, who insisted that the new discoveries far surpassed the limited scope of the ancients' knowledge.[6] Hariot's declaration that "seeing" was a privileged source of knowledge clearly placed him among the moderns in the debate. Indeed, Hariot's mathematical achievements were later cited by the modern advocate George Hakewill in his *An Apologie or Declaration of the Power and Providence of God* as evidence of new inventions unknown to the ancients. Conversely, Richard Harvey and Thomas Nashe, personal enemies united only by their suspicion of the new knowledge, refer to Hariot as a materialist and subversive, undermining the traditional Christian worldview.[7]

Although the sources of controversy may be debated, the single event that contributed most to this epistemological crisis was undoubtedly the discovery of America.[8] The sheer magnitude of the discovery, which was clearly not covered by any of the canonical texts, cast a doubt on their reliability as sources of truth. This did not mean that the canon was suddenly cast aside in favor of unmediated experience. J. H. Elliott shows convincingly that the impact of the early voyages on the period's scholars was relatively mute, and that they persistently attempted to maintain the authority of the canon.[9] The German humanist Joannes Cochlaeus was probably exceptional in his zeal to preserve the established geography and deny the significance of the new discoveries. In an introduction to Pomponius Mela's *Cosmographia* he insisted that whether the voyagers' reports were true or false "it has nothing . . . to do with Cosmography or the knowledge of History. For the peoples and places of that continent are unknown and unnamed to us. . . . Therefore it is of no interest to geographers at all."[10] But even more moderate scholars, who did not reject the discoveries outright like Cochlaeus, sought various ways of integrating the New World into their established worldview.

A fine example of the inner tension between a system of knowledge based on textual hermeneutics and a set of unprecedented new "facts" can be found in Anthony Pagden's study of Bartolomé de Las Casas's *History of the Indies*.[11] According to Pagden, despite spending most of his life in the New World, Las Casas was not satisfied to report his own experiences in the new lands. The ultimate source of knowledge for Las Casas lay in the established authoritative texts. He therefore constantly attempted to anchor his history in the established canon by quoting from it extensively and rather indiscriminately.

Las Casas tells the reader that he was moved to study the condition of the Indians not simply by observing their suffering at the hands of the colonists, but by reading a passage in *Ecclesiasticus*. These readings led to observations, which in turn led him to further readings in the canon. Las Casas, as Pagden points out, begins with a text and ends with more texts, while his own observations serve merely as a bridge between the canonical sources. When Las Casas does, inevitably, draw on his personal experiences as evidence for his claims, he does so reluctantly, well aware of the precariousness of his claims. Personal experience, after all, was an unreliable source of truth in early modern Europe.

Hariot's position clearly went beyond the uneasy compromise that Las

Casas attempted to maintain between new facts and ancient authority. If, as Hariot wrote, "the truth when it is seen is knowne without other evidence," then one can rely solely on eyewitnessing and completely dispense with traditional authority. Where Las Casas attempted to integrate his experiences into the traditional framework, Hariot suggested that personal experience was a satisfactory source of true knowledge. The act of seeing, in his view, was the privileged source of all truth, while other sources, such as authority or logic, were relegated to a secondary role.

Hariot's own treatise on the New World is therefore very different from Las Casas's accounts. The classical and biblical references are completely absent from Hariot's *Briefe and True Report of the New Found Land of Virginia*.[12] Instead, Hariot insists that his account is based entirely on his personal observations. In his introduction he describes himself as "one that have beene in the discoverie" and having "seene and knowne more than the ordinarie."[13] Conversely, the detractors of the Virginia colony are described by Hariot as those who "have spoken of more than ever they saw."[14] He then proceeds to list exactly what it was that he saw in Virginia: the plants, the minerals, the fauna, and the people. All descriptions are authorized solely by his own status as an eyewitness.

Other English travelers of Hariot's day clearly shared his convictions on the proper way to knowledge. While John Davis's *The Worldes' Hydrographcal Description* of 1595 stated its intention to rely on the "Aucthoritie of Writers" as well as "experience of Travelers" and Robert Hues's *Tractatus de Globis* of 1592 quoted extensively from the ancient sources, many other geographical treatises relied solely on personal experience.[15] The various accounts of Frobisher's expeditions in search of the Northwest Passage (1577–80) were all written by participants who recorded their own experiences. Walter Raleigh, Hariot's patron, used the same approach in his reports of his doings in Guyana: *The Discoverie of Guiana* of 1596 is a first-person account of his voyage in search of El Dorado, dependent solely on Raleigh's credibility as an eyewitness. Lawrence Keymis, Raleigh's lieutenant, employed the same strategy in his report on his exploration of the coast of Guyana in 1596. His testimony that the shoreline is riddled with rivers leading to the interior is based on nothing more than his claim that he had seen the rivers himself.[16] Whereas Las Casas felt a constant need to refer back to authoritative written sources, Raleigh and his circle were content to stake their claim to knowledge on "seeing" alone.

Hariot's and Raleigh's image of knowledge seems to us natural and unproblematic. We are, after all, epistemological heirs to Baconian empiricism and Cartesian rationalism, both of which regarded "accepted authority" as a source of error rather than truth. Raleigh's own experience, however, demonstrates how radical this view was in his own time and how precarious was the position of those who adhered to it. On his return from his voyage to Guyana in 1595, Raleigh announced the discovery of the hidden kingdom of El Dorado and a huge gold mine. Since Raleigh's statement was based on nothing more than his own testimony, the claim was viewed as inherently unreliable. Rumors circulated that Raleigh never really sailed to the New World, but spent his time hiding in Cornwall or privateering off the Barbary Coast.

After fifteen years of imprisonment due to his suspected subversive activities, Raleigh chose to stake his life and freedom on the truth of his report. In 1617 he was granted permission by King James to make one more voyage to Guyana and bring back evidence of the existence of the mine. The voyage ended in disaster: Raleigh's son was killed in a skirmish with the Spaniards, and Lawrence Keymis, who Raleigh blamed for the incident, committed suicide aboard ship. Raleigh returned to England and, failing to produce the required evidence, was executed the following year. Raleigh paid the ultimate price for his attempt—and failure—to establish a truth claim by relying solely on his personal experience. Basing true knowledge on mere unmediated observation was clearly a risky proposition in early modern England.[17]

Hariot was not deterred by his patron's misfortunes: despite the risks, he followed his maxim consistently in his scientific work as well as in his American writings. In the *Briefe and True Report* he insisted that he was reporting simply what he himself had observed. In his scientific work on the structure of matter, the mechanics of refraction, and the composition of the mathematical continuum, he followed the very same strategy. All these investigations were guided by his unflinching desire to actually see the subject matter and report his observations.

Hariot, of course, was faced with serious obstacles in his ambitious venture. "Seeing" into the structure of matter is a different matter than "seeing" the landscape of Virginia. It is an extremely difficult proposition, to put it mildly, especially given the limitations of Tudor and Stuart technology. The difficulties of "seeing" the composition of the continuum are even greater, as they are not merely technical, but conceptual. An attempt to "observe" a mathematical entity implies a visual, if not material, conception of mathematical objects. Hariot was implicitly suggesting that mathematical truth could be arrived at by close physical observation, rather than abstract and rigorous logical deduction. The project was ambitious indeed, and it grew directly out of Hariot's epistemology. Since observation was, for Hariot, the ultimate source of knowledge, he attempted to apply it even in seemingly unpromising fields.

In this chapter and the following one I explore the ways in which Hariot applied his "seeing" creed to his various activities. Beginning with his writings on the New World, I demonstrate that Hariot relied on personal observation as a privileged source of authority. The implicit claim of his *Briefe and True Report* was that what Hariot himself saw in America was being reproduced in writing for the benefit of those who stayed home. Hariot sought to make his readers "virtual witnesses" to things he himself had observed overseas.[18] In reviewing these texts, it will become apparent that although Hariot claimed the status of an innocent observer, a mere receptacle for the sights of the New World, his actual accounts were far from innocent. In fact, as I argued in Chapter 2, Hariot's exploration texts are profoundly shaped by a familiar narrative shared by his social circle—the standard narrative of exploration.[19]

Moving from exploration texts to Hariot's optical theories, I argue that the same desire to "see," which was evident in his accounts of the New World, is also at work here. Hariot seeks to actually view how a ray of light passes through a medium. In tracing the ray's passage, he draws on his experience as a geographical explorer and reconstructs optical refraction as a voyage of discovery. In the process, the optical medium is reconfigured as an undiscovered country and acquires many of the characteristics of the newly found lands of contemporary travel narratives.

In the following chapter I show how Hariot structured his mathematical approach according to the same principles. Clearly, he was not content with the traditional view of mathematics as a set of abstract logical relations. True knowledge of mathematics, for Hariot, could only result from actually "seeing" the structure of its operations. Specifically, Hariot sought an unmediated view of the mathematical continuum, perceived as a real, possibly material, object. Like the material medium of his optics, Hariot's mathematical continuum also came to resemble his own and his colleagues' descriptions of the shores of America.

"The truth when it is seen is knowne without other evidence," Hariot wrote, but there is no way to determine the context that inspired him to such musings. He may have been referring to his scientific work, proclaiming the primacy of direct observation over established authority. He may, just the same, have been referring to Raleigh's predicament, insisting that his patron's claims on Guyana should be credited even in the absence of supporting evidence. There is no way of knowing whether the context was geographical exploration or scientific investigation. In fact, it does not really matter. Hariot adopted this explorers' creed of direct observation and applied it to his scientific work. Whatever the subject of his investigation, Hariot always treated it as a voyage of discovery, designed to make him and his readers eyewitnesses to the inner workings of nature and mathematics. He sought to explore his scientific subject matter in the same manner in which he explored Virginia. In the process, the objects of his investigations came to closely resemble newly explored geographical lands.

SEEING NEW LANDS

In 1585, as part of the preparation for the expedition to Virginia sponsored by Walter Raleigh, Richard Hakluyt the Elder published a short pamphlet that he called "Inducements to the Liking of the Voyage Intended towards Virginia."[20] Hakluyt began by enumerating the various advantages that will accrue to the realm as a result of the expedition, ranging from the Christianizing of the natives to a long list of commercial benefits. Hakluyt then proceeded to carefully list the products that were to be sought in the new land: "Flaxe, Hempe, Pitch, Tarre, Masts, Clap-boord, Wainscot, or such like," which were previously purchased in the Baltic, may now be found in Virginia, he predicts optimistically.[21] Furthermore, "the mines there of Golde, of Silver, Copper, Yron, &c.," the "rich soile there for graine . . . Vines there for Wine, . . . Olives for Oile; Orenge trees, Limons, Figs," and so on, will provide employment for "our people void of sufficient trades."[22]

In order to achieve these lofty goals, Hakluyt assigned a crucial task to Raleigh's colonists: "The soile and climate first is to be considered, and you are with Argus eies to see what commoditie by industrie of man you are able to make it to yeeld, that England doth want or doth desire."[23] The first stage, then, is deliberate and careful observation. The colonists must carefully scan the land and later report their findings to an expectant England.

While Hakluyt's suggestion was here couched in general terms, it later became apparent that he had a particular observer in mind who was charged with the task of spying on the land with "Argus eies." In 1587, a year after the return of the colonists to England, Hakluyt dedicated a translation of an account of the French colonizing efforts in Florida to Walter Raleigh: "The particular commodities [of Virginia] . . . are well known unto your selfe and some fewe others, and are faithfully and with great judgement committed to writing, as you are not ignorant, by one of your followers, which remained there a full twelvemonth with your worshipful lieutenant M. Ralph Lane in the diligent search of the secrets of those countries."[24] Raleigh's follower, who stayed in Virginia a whole year with Ralph Lane and later reported his findings to Raleigh was, of course, Thomas Hariot.[25]

Hakluyt knew Hariot well and mentions him several times in his writings. In a dedication to his 1587 edition of Peter Martyr's *Decades*, for example, he congratulated Raleigh for promoting the mathematical sciences and having "maintained in your household Thomas Hariot, a man preeminent in those studies" so that Raleigh and his captains "by his aid you may acquire those noble sciences." Many years later, while advocating a renewed colonizing attempt in Virginia, he cited "Master Thomas Heriot, a man of much judgement in these causes" as an authority on the presence of gold in the interior.[26] When Hakluyt proposed his program for surveying the land of Virginia, he undoubtedly considered the young mathematician to be the most qualified person to carry out such a mission. It is no coincidence, then, that when Hariot, on his return, published his account of the new land, he structured it as a direct response to Hakluyt's pamphlet.

Hakluyt had proposed a list of commodities that the land may yield and which should be looked for by the colonists. Seafaring necessities, such as "Flaxe, Hempe, Pitch, Tarre, Masts," and so on, were high on his list, and he supplemented them with other desirables such as "Golde, Silver, Copper, Yron," and various agricultural products.[27] Hariot, responding directly to Hakluyt's challenge, configured his *Briefe and True Report* in a similar form: most of the *Report* is structured simply as a list of commodities that Hariot had found in Virginia, with a single paragraph of explanatory text accorded to each. In his chapter "Of Marchantable Commodities" he lists "Silke of grasse or grasse Silke," "Worme Silke," "Flaxe and Hempe," "Allum," "Wapeih," "Pitch, Tarre, Rozen and Turpentine," "Sassafras," "Cedar," and later "Iron," "Copper," "Pearle," and "Sweete Gummes."[28] In a section "Of Fruites," he counters Hakluyt's hypothetical agricultural products with his own list of "Walnuts," "Medlars," "Metaquesunnauk," "Grapes," "Straberries," "Mulberries," "Sacquenummener."[29] He simply substitutes the actual commodities he had found in Virginia for the hypothetical list provided by Hakluyt.[30] Hariot, in fact, precisely fulfilled the mission as it was defined by Hakluyt: he scanned the country carefully, constantly noting and registering the various useful commodities that the land may be made to "yield" and, on his return to England, published a *True Report* based on these observations. He was, in other words, Raleigh's and Hakluyt's "Argus eies" in Virginia, eager to pass on to his senders everything he saw.

Hariot's status as the "eye" of the expedition is central to the narrative strategy of the *Briefe and True Report*. It serves him well, for instance, when in the preface he attempts to counter criticism from disenchanted former members of the colony. Their claims should not be believed, he argues, since they had not really seen the land: "The cause of their ignorance was, in that they were of that many that were never out of the iland where wee were seated . . . during the time of our abode in the countrey."[31] Hariot himself, in contrast, is not only "one that have beene in the discoverie," but, among the colonists, one that has "seene and knowne more than the ordinarie."[32] "Seeing," then, is not only Hariot's stated mission but also a source of knowledge and authority when confronted with the unfavorable reports on the colony. The *Briefe and True Report*, he implies, is an accurate depiction of overseas realities based on careful observation. It is, therefore, superior to the accounts of his adversaries, which do not have the benefit of such extensive "seeing."

Since Hariot was the "eye," his *Report* was simply a "view"—a transparent window, communicating what he saw overseas to the readers who stayed behind. He refers to his text in precisely these terms when, addressing potential sponsors of the project, he suggests that "you seeing and knowing the continuance of the action by the view hereof you may generally know & learne what the countrey is."[33] The implication is that the *Report* is a simple reflection of overseas realities and is, therefore, an acceptable substitute for actual travel. The reader of the text partakes in Hariot's observations and sees the world through his eyes.

But although Hariot claimed that his account was a transparent vehicle through which the "view" of overseas lands was transported to Europe, we need not take him at his word. English travel accounts of this period followed a familiar standard pattern, which was quite independent of the particular realities of the land explored. Although many travel narratives, like the *Briefe and True Report*, claimed to be no more than simple, unmediated accounts, they nonetheless shared a standard vision of America and of the explorers' role there. This story, which I refer to as the *standard narrative of exploration*, posited a rich golden land deep in the interior, which served as the ultimate goal of exploration. Unfortunately, the land was protected by seemingly insurmountable barriers: high mountains, forests, and other natural obstacles barred the way of the adventurous traveler. The explorers—usually the author of the tract and his companions—did not despair: the careful observation of the barriers revealed certain hidden passages that broke right through them and led the way to the golden kingdom of the interior.

Various characteristics of Hariot's treatise suggest that the master narrative of exploration is present even in his sober recounting of his observations in Virginia. There is no question, of course, that Hariot adhered closely to the features of the land that he had observed and gave a lucid account of his findings. Personal observation was, after all, Hariot's sole resource for endowing his tract with authority and credibility, and his claim that he was providing a transparent window opening out on the new land had to be sustained. But beneath the formal, unassuming surface of the tract, the basic outlines of the standard tale can be discerned.

First of all, it should be noted that although most of the *Report* is written in a steady, unassuming, factual tone, a certain uncharacteristic exuberance emerges occasionally, as if the mythical underpinnings of the text are showing through the placid surface. In describing, for example, the Indians' agricultural practices, Hariot claims that a single acre of land would yield 200 bushels of crops in Virginia, but only 40 in England.[34] The fertility of a tract of land in Virginia is, according to Hariot, five times that of an equivalent tract in England.

This idealized depiction would in itself point to the presence of the mythical undercurrent in the text. Hariot, however, is not satisfied even with this optimistic estimate of the richness. He goes on to claim that in Virginia one day's work on a tract of land of only 25 yards square would suffice to support and nourish a man for an entire year![35] If Hariot's previous assertion seemed overly optimistic, his claim here is clearly mythical. Virginia is reconfigured as an unspoiled Eden—a virgin land that will yield its produce to the colonizers with hardly any work or effort on their part. The outlines of certain fundamental characteristics of the standard narrative of exploration are clearly visible in this passage.

Even more important than the select mythical passages in the *Briefe and True Report* is the general plan of the treatise. The long list of products, which takes up much of the tract, serves to promote a vision of Virginia as a golden land of plenty. The sheer length of Hariot's list of valuable commodities, which the land "yields," offers a view of the country that goes well beyond the sum total of the separate items. The endless, monotonous, list of commodities, which follow each other in predictable unrelenting regularity, suggests a land whose wealth is unlimited. Each commodity is, undoubtedly, valuable in itself, and their combination even more so. The actual impression of unlimited wealth, however, is created not by the particular items, but by the list itself: each commodity is followed by another, and another, and another after that, in endless succession. The list seems to have no end, and the riches of Virginia have no limit.

In the conclusion it indeed becomes clear that Hariot never intended the list to be finite. If his readers yet remain skeptical of the wondrous wealth of Virginia, he argues, that is merely because only a small part of the land has been explored. What had already been found, he declares, "is nothing to that which remaineth to be discovered." The further one proceeds inland from the sea coast, the richer, more fruitful, and more wondrous the land becomes.[36]

This is a familiar ruse of the standard narrative of exploration, which invariably places the golden land beyond the immediate reach of explorers. Equally important, however, is the fact that the promised riches of the interior offer Hariot an unlimited supply of commodities to add to his already lengthy lists. Whatever is yet lacking in the lists of the *Report*, he implies, will surely be included after further exploration of the interior. The *Briefe and True Report*, then, is only a beginning: more and more items will eventually be appended to it as the exploration and discovery of Virginia proceed. The land of Virginia that Thomas Hariot presents to his readers is, in the end, the golden land of the standard narrative of exploration. Despite the sober tone governing much of the text, the Virginia of the *Briefe and True Report* comes off as a mythical land of unlimited plenty.

The text of the *Briefe and True Report* presents the vision of a glorious land of great riches and suggests the theme of a wondrous interior. John White's map of "The Arrival of the Englishemen in Virginia," appended to de Bry's 1590 edition of the *Report*, and Hariot's notes to it, present the narrative of exploration in its most complete form (see Chapter 2, Figure 3).[37] The rich land of the interior, the natural obstacles protecting it, and the persistent explorers who scour the coast for a passage before finally breaking through to the promised land are all accounted for in this map. The master narrative of exploration was a persistent undercurrent in the main text of the *Briefe and True Report*. In the "Arrival" map the standard tale comes clearly to the fore and shapes the geography of the land.

Hariot's account of Virginia is informed by the tension between two conflicting themes. On the one hand is the desire to "see" everything in the new land and to pass on the findings clearly and accurately to those who stayed behind. As Hariot himself states, his intention in the *Report* was to open a window that would offer his readers an unmediated view of the realities of America. On the other hand, the wish to "see" the land became inseparable from a set of preconceived expectations of what will, in fact, be discovered. Although Hariot promises a simple, straightforward account of his findings, his text is in fact imbued with a mythical tale—the standard narrative of exploration. This tale supplied structure and meaning to the explorers' doings in Virginia and provided an attractive framework through which the readers could interpret Hariot's account. At the same time, the all-too-familiar legendary themes tended to undermine the text's claim to be an unmediated report and implicitly questioned its credibility.

The tension between the two trends inherent in Hariot's exploration writings reached its highest pitch during his involvement in Raleigh's search for El Dorado. Hariot himself was not a member of the expeditions to Guyana of 1595-96. His information was based solely on Raleigh's and Keymis's accounts of Guyana, and he treated them just as he treated his own report on Virginia—as direct, unmediated representations of a distant reality. His unquestioning attitude is manifest in his letter to Sir Robert Cecil dated July 11, 1596, a short while after Keymis's return from Guyana:

Right Honorable Sir,

These are to let you understand that whereas according to your honors direction I have been framing of a Charte out of some such of Sir Walters notes & writinges which he hath left behind him, his principall Charte being carried with him. If it may please you I do thinke most fit that the discovery of Captain Kemish be added in his due place before I finish it. It is of importance, & all Chartes which had that coast before be very imperfecte in many thinges elce. And that of Sir Walters although it were better in that parte then any other, yet it was done but by intelligence from the Indians, and this voyage was especially for the discovery of the same; which is, as I find, well and sufficiently performed.[38]

Keymis and Raleigh are both described in the letter as competent observers who reliably recorded what they had seen and passed their reports on to Hariot, who in turn sets them out in the form of a map. Although Hariot himself is not the actual observer in this instance, "seeing" is, nonetheless, the main purpose of the expedition. Knowledge based on other sources, such as "intelligence from the Indians," is considered suspect.

Hariot's sustained attempt to base his account solely on unmediated seeing, however, shows signs of strain when he moves to discuss El Dorado, the ultimate goal of the expedition. The search for El Dorado was, after all, the purest example of the myth exploration.[39] In discussing it, Hariot was forced to compromise his reliance on firsthand observation: "Concerning Eldorado which hath been showed your honor out of the Spanish booke of Acosta which you had from Wright & I . . . I shall shew you it is not ours that we meane there beinge three. Nether doth he say or meane that Amazones river & Orinoco is all one . . . as by good proof out of that booke alone I can make manifest."[40] Hariot here goes well beyond anything that was seen by Raleigh and Keymis. His knowledge of El Dorado is based solely on a written source—Jose de Acosta's *Historia Natural y Moral de Las Indias* of 1590. Furthermore, in drawing specific conclusions about the geography of the land, he falls back on a traditional source of knowledge, namely, the interpretation of a text: the separateness of the Orinoco and Amazon can be proven, according to Hariot, by "that booke alone." Although in the rest of the letter Hariot is scrupulously faithful to his belief in the primacy of personal observation, in his discussion of El Dorado he relies on an older order of knowledge—hermeneutics and textual commentary.

The Raleighan version of the myth of El Dorado was superbly expressed in a poem by George Chapman, a prominent member of Raleigh's circle and a strong supporter of his colonizing ventures. When Lawrence Keymis published his account on the *Second Voyage to Guiana* in 1596, Chapman contributed a poem, "De Guiana Carmen Epicum," which prefaced Keymis's report. A few verses should suffice to establish the basic tenor of Chapman's lengthy piece:

> Riches, and conquest, and renown I sing,
> Riches with honour, conquest without blood,
> Enough to seat the monarchy of earth,
> Like to Jove's eagle, on Eliza's hand.
> Guiana, whose rich feet are mines of gold
> Whose forehead knocks against the roof of stars,
> Stands on her tip-toes at fair England looking,
> Kissing her hand, bowing her mighty breast,
> And every sign of all submission making,
> To be her sister, and the daughter both
> Of our most sacred Maid: whose Barrenness
> Is the true fruit of virtue, that may get,
> Bear and bring forth anew in all perfection,
> What heretofore savage corruption held
> In barbarous chaos; and in this affair
> Become her father mother and her heir.
>   . . .
> But you patrician spirits that refine
> Your flesh to fire, and issue like a flame
> On brave endeavors, knowing that in them
> the tract of heaven in morn-like glory opens,
> that know you cannot be the kings of earth
> (claiming the rights of your creation)
> And let the mines of earth be kings of you;
> That are so far from doubting likely drifts,
> that in things hardest y'are most confident;
>   . . .
> But where the sea in envy of your reign
> Closeth her womb as fast as 'tis disclos'd,
> that she like avarice might swallow all,
> and let none find right passage through her rage;
>   . . .
> But we shall forth I know; gold is our fate,
> Which all our acts doth fashion and create.
>   . . .
> And now a wind as forward as their spirits,
> Sets their glad feet on Smooth Guiana's breast,
> Where (as if each man were an Orpheus)

A world of savadges fall tame before them,
Storing their theft free treasuries with gold,
And there doth plenty crown their wealthy fields.[41]

Chapman's understanding of the search for El Dorado closely follows the views expressed by the explorers themselves. Guyana is seen as withholding "riches" and "mines of gold," thus presenting a challenge to the "patrician spirits" bent on "brave endeavors," which they cannot refuse. The way, however, is blocked by formidable obstacles, foremost among them is the sea, which would "let none find right passage through her rage." The explorers, nonetheless, proceed undeterred: "But we shall forth I know; gold is our fate, / Which all our acts doth fashion and create," Chapman announces confidently. His faith is then rewarded when "A world of savadges fall tame before them, / Storing their theft free treasuries with gold." All the elements of the familiar narrative of exploration are firmly in place in Chapman's poem: the hidden treasures, the obstacles along the way, and the enterprising explorers who, despite all difficulties, find their way to the golden land where they are rewarded with great riches. The search for El Dorado is the incarnation of the myth of exploration in its purest form.

The actual map of Guyana that Hariot produced on the basis of his varied sources attests to the inner tension in his work (see Chapter 2, Figure 6). Parts of the map clearly reflect the detailed information that Hariot acquired from Raleigh and Keymis. The general outlines of the coast, the position of the island of Trinidad in relation to the Orinoco, and the names of the rivers cutting through the shoreline are all indicative of Hariot's access to firsthand knowledge of the area. The same impression can be gleaned from a sketch of Trinidad and the Guyana coast, found in Hariot's papers, which attests to his diligence in transforming the explorers' accounts into cartographic form.[42]

The map as a whole, nonetheless, openly conveys the tale of exploration and discovery. The golden land in the interior, on the shores of a lake, surrounded by mountains but potentially accessible through winding rivers, is the epitome of this familiar myth. Like Hariot's letter to Cecil, it reflects the tension between the guiding themes of Hariot's exploration work. On the one hand, he openly insists that intelligence on the land could only be gained through personal observation; on the other, like many of his colleagues in the exploration enterprise, Hariot had specific expectations of what this intelligence would be. Paradoxically, although knowledge could only be gained by seeing, what would be seen was largely predetermined.

The picture becomes even more complex if we consider that the two sides of the paradox are not truly separate, but interdependent. After all, the stated motive for the "seeing" expeditions was precisely the discovery of the mythical land. The sights that were so carefully observed and recorded were chosen for their usefulness as signposts on the way to the golden interior. Thus, for example, Keymis carefully explored the rivers along the coast of Guyana precisely because he considered them potential passageways to the "plaine of Empire." John White and Thomas Hariot showed a similar interest in the Carolina coastline for the same reason—they considered it a gateway to a wondrous land. In this way, the myth of exploration profoundly affected even the careful and meticulous observations of the New World.

On the reverse side, the overwhelming desire to "see" shaped the tale of exploration itself in crucial ways. The inevitable passages carved through the land, which are a fundamental part of the myth, are nothing if not avenues for seeing. They are put in place precisely to provide pathways for the enterprising explorer bent on viewing everything. The passages to the golden interior, in other words, incorporate the all-seeing eye of the explorer into the center of the tale of exploration. The desire for exact unmediated recording of observed fact is thus made an integral part of the myth of the golden land.

The two sides of the paradox are, therefore, inseparable. The desire for precise observation and recording is dependent on faith in the myth of the golden land, while that very tale takes into account the observing explorer. The result is the peculiar depiction of the geography of America discussed above in Chapter 2: exact, carefully measured coastal outlines are made an integral part of purely legendary depictions. The correct depiction of the locations of Frobisher's and Hudson's "straits" in James Beare's map, for example, is part of a chart of a nonexistent northwest passage.[43] Similarly, Hariot's modern-looking outlines of the coast of Guyana and Virginia are incorporated into the wider story of exploration and discovery. In all of these cases, the authors do not distinguish between what was actually observed and measured and what was based on legend. The maps and accounts flow unobtrusively from one to the other. For Elizabethan mapmakers and explorers, the desire to see and record, and the myth of the golden interior, were mutually supportive components of a single vision.

Thomas Hariot carried this paradoxical approach beyond the confines of the discourse of geographical discovery. Other mathematical practitioners adopted the rhetoric of exploration in describing their own technical work, but Hariot went much further.[44] His novel theory of optical refraction, his views on the composition of matter, and his reconstruction of the mathematical continuum were all structured along the pattern set by the discourse of exploration. As in the case of Guyana and Virginia, Hariot was driven by an unrelenting desire to "see" his subject matter. Although this proved to be impossible in most cases, Hariot persisted undeterred: he proceeded to reconstruct the elusive subject matter in accordance with his predetermined notions of what it "should" look like. Hariot, in other words, constructed a visual map of unseen regions, just as he did in his depictions of the New World. It should come as no surprise that his charts of the invisible realms of microscopic matter and the abstract mathematical continuum closely resembled his maps of the lands overseas.

SEEING MATTER: HARIOT'S THEORY OF REFRACTION

In the introduction to his *Achilles' Shield*, George Chapman included a poem dedicated "To My Admired and Soule-Loved Friend, Mayster of all essential and true knowledge, *M. Harriots*." Early in the poem Chapman asks his learned friend's assistance in gaining insight into his own troubled soul:

> O had your perfect eye Organs to pierce
> Into that Chaos whence that stiffled verse
> By violence breaks: where Gloweworme like doth shine
> In night of sorrow this hid soule of mine.[45]

Chapman's association of Hariot with vision and observation is, of course, no coincidence. Hariot is here described as possessing a "perfect eye," capable of piercing through into the "Chaos" that lies within Chapman's soul. Already we are reminded of the "Argus eies" attributed to Hariot by Hakluyt on the eve of his voyage to Virginia. His role in both instances is the same: to penetrate the interior and report his findings.

It is clear from the wording of this verse that Chapman does not really expect Hariot to be able to comply with his request. Chapman's soul is his own private domain, inaccessible to anyone but himself. The significance of the verse lies in its generalized description of what Hariot excels at, in Chapman's view: his role is to observe and penetrate, and the fact that he is unable to do so merely accentuates the depth of Chapman's suffering. Later in the poem, when Chapman turns from his own troubles to the praise of his friend, he explicitly states his views on Hariot's achievements:

> And when thy writings that now errors Night
> Chokes earth with mistes, breake forth like easterne light,
> Showing to every comprehensive eye,
> High fectious brawles becalmed by unitie,
> Nature made all transparent, and her hart
> Gripte in thy hand, crushing digested Art.[46]

The secrets of nature, in Chapman's view, are hidden behind the "mistes" of "errors Night." Hariot's writings break through these obstacles like "easterne light," presenting what was previously hidden "to every comprehensive eye," bringing "unitie" and order to chaos, and power and glory to Hariot himself.

The wording and the rhetoric of Chapman's description are highly reminiscent of the rhetoric of exploration and discovery. The depiction of light breaking through mists and darkness seems as if it were lifted directly from the participants' accounts of Frobisher's voyages in search of the Northwest Passage. Dionise Settle, for example, reporting on his experiences in the expedition, described how "through darknesse and obscuritie of the foggie mist we were almost run on rocks and Islands. . . . But God (even miraculously) opening the fogges that we might see clearly."[47] On a different occasion he described how "the fogges brake up and dispersed so that we might plainly and clearly behold the pleasant ayre, which so long has been taken from us, by the obscuritie of the foggie mistes."[48] Hariot's attempts to penetrate the secrets of nature is not so different then, in Chapman's view, from the efforts of geographical explorers to break through physical fogs and obstacles.

If some images employed by Chapman seem to recall the hazards of Arctic exploration, others seem to refer to the more immediate past of the Guyana voyages, in which both Chapman and Hariot were involved. Indeed, the imagery of this poem is often strikingly similar to Chapman's own *De Guiana* of two years earlier. Addressing Hariot, for instance, Chapman calls on him to "Pierce into that Chaos" that is his soul, just as he breaks through to nature's secret, quieting "high fectious brawles" with "unitie." In *De Guiana* Queen Elizabeth is called on to

> Beare and bring foorth anew in all perfection,
> what heretofore savage corruption held
> in barbarous Chaos

Chapman goes on to predict a similar happy fate for Raleigh's followers in Guyana: "A world of savadges fall tame before them,/Storing their theft-free treasuries with golde."[49] The similarity between the passages in the two poems is unmistakable: Hariot brings unity to chaos and tames "high fectious brawles," just like Elizabeth and Raleigh bring peace and prosperity to the barbarous world of the Guyanan savages.

Beyond the literal similarities and the tropes shared by the two poems lies a deeper connection: Chapman's description of Hariot, his glorification of Raleigh's conquistadors, and even Settle's account of the Canadian arctic, are all part of the same discourse and share the same narrative: the standard tale of exploration, which posits hidden secrets in the interior and promises great riches to the traveler who breaks through, is common to all sources. In Chapman's view, Hariot the scientist is no different than Raleigh the explorer, piercing through mists to reveal stable truths.

Chapman's poem combines both sides of the paradox of "seeing" that we have already observed in Hariot's discussions of Virginia and Guyana. On the one hand, Hariot's role is clearly to observe and record whatever he may encounter. After all, he is described as having a "perfect eye," and his goal is to make his findings known "to every comprehensive eye." On the other hand, Hariot clearly knows what he will find and what effect he will have on his discoveries—"High fectious brawles becalmed by unitie." Just as was the case in his geographical accounts, the explorer of the unknown is well aware beforehand what will be found, since the nature of the discovery is foretold in the standard narrative of discovery. There is a distinct tension in Chapman's poem between Hariot's role as explorer of the unknown and the fact that, like English explorers overseas, he fully expected his findings to accord with his predetermined notions.

This tension becomes of more than literary interest when Hariot translates these generalized views of his role into specific scientific views. Hariot's theory of the nature of light and the causes of refraction was shaped to a large extent by the very rhetoric of the scientist/explorer discussed above. At the heart of his theory was Hariot's desire to literally "see" the passage of light rays through a material medium. At the same time, he never doubted that he already knew what this process really looked like from close up. The result was a unique theory of refraction, which placed the standard tale of exploration within the inner workings of nature.

On October 2, 1606, Johannes Kepler wrote a letter to Thomas Hariot, asking for his views on various scientific matters. The two were not previously acquainted, and although Hariot made it clear during the ensuing correspondence that he was familiar with Kepler's work, the reverse was apparently not the case:

> Joannes Eriksen, most excellent Hariot, the bearer of these letters, caused
> me great joy when he told me there is a man in England most versed in all
> the secrets of nature; who if not impeded, desired to correspond with me
> through letters; who possesses, especially in optics, new principles, unknown
> to the public, from which both my Optics book, and whatever was produced
> before, are found to be not only deficient, but even erroneous.[50]

Kepler was indeed not far from the mark. Throughout his career Hariot continually engaged in optical experimentation and theorizing and had impressive results to show for his efforts.

Kepler had apparently heard a great deal about Hariot's views from their mutual friend Eriksen. In his letter, he proceeded to pose questions to his correspondent on a wide variety of topics, ranging from the theory of weights to Hariot's views on astrology. Much of the letter was, however, devoted to optics. What is the origin of colors and what are the essential differences between them, Kepler inquired. Why do colors appear when white light is refracted, and what accounts for the height of the rainbow?

Despite his supposed eagerness to correspond with the Imperial Astronomer, Hariot took his time in responding to Kepler's queries. When he did eventually answer two months later, he declined to answer all of Kepler's queries, claiming that even "to respond to a single one in one place is too much, especially in such a short time as my answer is expected."[51] Instead, Hariot chose to focus on a single optical question—the problem of refraction. Whereas Kepler's inquiries on this subject were merely part of a long series of questions, Hariot makes it clear that for him this question is of special importance.

Refraction is the change in the direction of a ray of light when it moves from one medium to another. The phenomenon had been well recognized since antiquity and was discussed at length by Ptolemy.[52] Although no quantitative rules were known before Hariot's time, the basic qualitative features of the phenomenon were generally accepted: light rays passing from one medium into a denser one bend closer to the perpendicular to the plane of the interface; conversely, in passing from a dense medium to a rarefied one, a ray is refracted further from the perpendicular. The importance of refraction to optical theorists since antiquity lay in its centrality to the study of two classical optical problems: the formation of the rainbow and the understanding of vision.[53] Our two correspondents were probably the most significant contributors to the study of refraction and its effects during the period: Kepler studied the formation of the rainbow extensively and was the first to formulate the theory of the retinal image. Hariot was probably the first to propose the mathematical law of sines that governs refraction, perhaps as early as 1597. He was also, most likely, the first to mathematically calculate the correct radius of the rainbow.[54] His studies of prisms and formation of colors fill many folios of his unpublished papers.[55]

In his letter, Hariot offered Kepler a table of refraction, one of many that can still be found among his papers. The table listed fifteen different substances serving as optical mediums, such as oil, wine, salt water, crystal, and glass. The density of each substance was estimated by its "weight ratio" when compared to fountain water of equal volume. The angle of incidence for all substances was listed as 30 degrees and was followed by the angle of refraction that Hariot had measured for each of the substances. Kepler, of course, had never asked for the table, but Hariot remained confident in the value of his reply: "This table tells you many things that can be explained in the length of your letter. Which I leave to your genious to speculate. Should it not be concluded that my letter equals yours?"[56]

Hariot's tone is outright arrogant, if not taunting, and it is difficult to say why he chose to adopt this attitude with the leading natural philosopher in Europe. He continued in the same vein, challenging not only Kepler but also the established optical tradition explaining the physical causes of refraction. Since Alhazen's *De Aspectibus* became known in the West in the thirteenth century, it had been commonplace to explain refraction by the physical resistance of a material medium to the ray passing through it. According to Alhazen and his Western followers—Roger Bacon, John Pecham, and Witelo, in particular—refraction was caused by mechanical friction: when a ray passing through a medium obliquely strikes a denser one, they argued, it encounters sudden resistance and is deflected toward a less resistant path—that is, closer to the perpendicular. Conversely, passing from a dense to a rare medium, a ray encounters less resistance, and consequently is deflected away from the perpendicular.[57]

According to Hariot, however, this traditional view is contradicted by the experimental results of his refraction table. Rays passing through dense mediums are sometimes refracted away from the perpendicular, while in rare mediums they can be refracted toward it, "contrary to ancient and popular laws."[58] He continues, "Do you not see that from crystal to amber, or from gum to crystal there is no refraction? But who doubts that these mediums differ widely in rarity and density? It is a paradox of optics, but true."[59] (See Figure 9.)

Before suggesting any solution to his paradox, Hariot first expounds on the difficulties:

> Of the reason for refraction at this time I relate nothing, before the following argument:
>   Let there be some luminous ray in a rare medium AB. It strikes a surface of some diaphanous density [medium] in point B. At that moment, what will the ray do? It will proceed, no? It will proceed, you say, because the medium was posited as diaphanous, and permeable to rays. It does not proceed, however, directly to point E, but is refracted and inclined in point B towards perpendicular BI, and penetrates to point C. This, I concede, experience teaches. But why? You say, it is impeded in its progress by the surface , and is therefore refracted. But is there not the same impediment after the entrance? Is not the surface of the material of the same nature as the entire body? Are not the same surfaces in the entire body in every position, even if they are not seen? Does it not, therefore, perpetually refract in the passage just as in the beginning?[60]

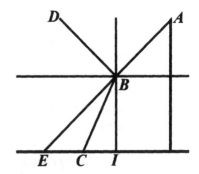

Figure 9. Hariot's sketch of refraction, accompanying his letter to Kepler of December 2, 1606. Printed in *Johannes Keplers Gesammelte Werke*, vol. 15 (Munich: Beck, 1951), 367.

Hariot's point seems well taken: if indeed refraction is caused by the resistance of the medium, why does it occur only at the surface of interaction? Why is the ray not continuously refracted by all the surfaces it encounters in its passage within the optical medium? He then proceeds to offer what he considers would be Kepler's likely response: "You answer: because the ray encounters no actual surface in the body, but such force, because it encounters as if infinite surfaces in every position, the ray does not know where to turn, and therefore, after the first refraction by the actual surface, it proceeds without other flexions directly to point C."[61] Kepler's supposed response, then, is that since the inner surfaces within the medium are positioned in all possible angles, their effects on the ray cancel each other out. Only the first external surface is unique, and therefore it is the only one to affect the direction of the ray.

Continuing his imagined dialogue with his supposed interlocutor, Hariot immediately challenges this position:

> I say that it is of greater convenience to the ray to proceed directly on the line BE in the beginning than by refraction to decline, as it cannot encounter a cause in the beginning, but get around it during the passage, that is the impediment of resistance. And since there would be the same resistance everywhere, it would be similarly deflected perpetually. But how so, and by what means? Expect and wait a little while. We return to the beginning where we started.[62]

Hariot's position is, then, that if physical resistance were the cause of refraction, then it would necessarily affect the ray throughout its passage, and not only at the beginning. Therefore, according to Hariot, an alternative explanation must be found.

Before returning with Hariot to "the beginning," it is worth noting that Kepler himself did not consider Hariot's objections to be particularly daunting. In his response dated August 2, 1607, he seems rather puzzled by Hariot's reasoning as well as by his combativeness:

> You construct an objection why a ray is not refracted in its entire passage, and posit me to respond. . . . In fact, I repudiated the consideration of both the end and the good in optical laws, and I gave a cause why surfaces refract and bodies do not. Light, I said, is of a genus which participates in surfaces, therefore it submits to surfaces but not to bodies, with which it does not participate. Read the first chapter of my optics.[63]

Kepler bridles at Hariot's tactic of assigning him imaginary positions in a reconstructed dialogue. For Kepler, the teleological position that Hariot attributes to him, that is, that the ray proceeds to point C because it does not "know" where else to turn, is clearly unacceptable. His objection to Hariot's explanation is even more fundamental: he simply refuses to acknowledge the problem posed by Hariot and is therefore uninterested in a solution based on its assumptions. Kepler is quite content to interpret light and its effects in traditional Aristotelian terms: "light" shares certain qualities with "surface" and is therefore affected by surfaces, while it shares no qualities with "corpulence" that therefore leaves it unaffected. When treated in formal Aristotelian terms, Hariot's "paradox" simply vanishes.

Hariot, of course, is not convinced by Kepler's reasoning: "If these assumptions and reasons satisfy you, I wonder," he wrote to Kepler on July 13, 1608.[64] He was clearly looking for an actual material explanation, and in those terms, he argues, refraction is extremely problematic:

> When the ray AB hit a surface of dense diaphanous (matter) in point B and proceeded refracted: I ask whether it was not at the same instant reflected from the same point B? That a second angle, equal to the incidence, is reflected to D is not to be doubted. Experience teaches this, the eyes see: but it remains dubious of the identity of time and points. Then no answer was given without absurdity. For the same point would be permeable and resistible to a ray at the same time and the same part of the ray would be in two places simultaneously.

The paradox of refraction is here stated unambiguously: How can a single point both refract and reflect a ray of light? For in the case of refraction, light is allowed into the material medium, while in reflection it is denied entrance. How then, Hariot asks, can a single point be both permeable and impenetrable to light at the same time? Furthermore, even if an explanation for this would be found, simultaneous reflection/refraction would mean that a light ray would be present in two locations at the same time—refracted into the medium, and reflected outside it. A paradox indeed.

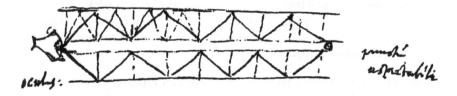

Figure 10. Thomas Hariot, "Transitus radiorum per medium" (The passage of a ray through a medium). British Library Add. MS 6789, f. 328 (Courtesy of the British Library).

Hariot does not leave his correspondent in suspense for long and proceeds directly to his solution of the difficulties:

> Which is therefore to be answered. Since one part of a contradiction is false, it necessarily follows that the other is true. It remains to be said that the same point is not both permeable and resistible to rays. Nor that the same ray is reflected from the surface of the dense diaphanous, and simultaneously received in the body. Therefore the dense diaphanous body which to the senses seems to be continuous in all its parts is in fact not so. But has corporeal parts which resist the rays, and incorporeal parts permeated by the rays. That is, refraction is nothing else but internal reflection, and the part of the rays received inside, and that seems to the senses to be straight, is then, in truth composed of many [parts].[65]

Hariot's meaning is made clear in his unpublished manuscripts. An undated diagram, entitled "Transitus radiorum per medium" (The passage of a ray through a medium), posits the eye and a "visible point" at opposite ends of two parallel tunnels (see Figure 10). A light ray is shown traversing each of these passages, reflecting back and forth from both sides of the tunnels until it reaches the eye. According to Hariot, then, the medium is permeated by openings, which provide passage right through it. Refraction, in this view, is indeed internal reflection, since it is the result of the rays being reflected from hard points within the passage. Furthermore, as Hariot says, the rays are not straight, but rather composed of many parts—the numerous segments that comprise its zigzag course.

Other diagrams in Hariot's papers indicate that he experimented with var-

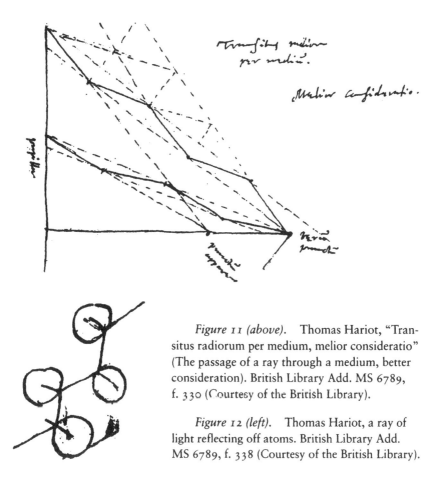

Figure 11 (above). Thomas Hariot, "Transitus radiorum per medium, melior consideratio" (The passage of a ray through a medium, better consideration). British Library Add. MS 6789, f. 330 (Courtesy of the British Library).

Figure 12 (left). Thomas Hariot, a ray of light reflecting off atoms. British Library Add. MS 6789, f. 338 (Courtesy of the British Library).

ious versions of this position. Another sketch of his reconstruction of refraction is entitled, appropriately, "Transitus radiorum per medium, melior consideratio" (that is, "better consideration") (see Figure 11). It shows the two rays proceeding from the "true point" to the eyes in diverging zigzagging courses, the "apparent point" being positioned at the intersection of lines extended from their angle of entry into the eye. Other sketches yet show close-ups of the rays of light bouncing back and forth between round physical atoms within the medium, or simply changing direction by contact with such particles (see Figure 12). An account of Hariot's atomistic view by his contemporary Nathaniel Torporley, the "Synopsis of the controversie of Atoms," also contains a diagram demonstrating Hariot's position.[66] In all cases, however, Hariot's general position is the same: the surface of the medium only appears to the eye to be smooth and homogeneous. On closer inspection it is revealed that it is traversed by passages that enable the ray of light to pass right through it.

Hariot's theory of refraction is clearly predicated on his desire to "see" the actual effect the material medium has on a ray of light. Kepler was quite satisfied to account for the phenomena in the abstract Aristotelian language of substances and qualities. He seemed content to base his views on their logical consistency and the authority of ancient philosophy. Hariot, in contrast, insisted that the actual physical passage of the rays be reconstructed and closely observed. For him, only actual "seeing" could be a satisfactory basis for an explanation. The close scrutiny of a supposedly homogeneous surface revealed inner passages through which light is seen to pass. When Hariot produced sketches of the passage of light through a medium, he was literally bringing into view the hidden workings of nature. "The truth when it is seen is known without other evidence," Hariot announced, declaring his belief in the primacy of vision as a source of knowledge. In his theory of refraction he applied this standard to his theory of light and vision itself (see Figure 13).

Despite the physical impossibility of tracing the passage of light through matter, Hariot does not hesitate to offer a detailed visual reconstruction. This, of course, is a rather paradoxical position for someone who insists on the importance of actual seeing. We have already noted a similar conflict in his travel writing: although Hariot always insisted on his status as a direct observer, he repeatedly constructed his newly discovered lands in accordance with a predetermined set of assumptions. His guide to the undiscovered lands was the standard narrative of exploration. It now becomes apparent that a similar tale served him equally well in his optical theory.

Hariot first encounters the problem of refraction as a confusing paradox: refracted light acts in unruly and confusing ways, which do not accord with traditional expectations. Furthermore, the established explanations of refraction turn out to be seriously flawed, and attempts to shore up their deficiencies prove unsuccessful. The smooth surface of the refracting medium is impenetrable to light rays, according to Hariot, and rebuffs any attempt to resolve the mystery of refraction. The solution to the paradox becomes obvious, however, once Hariot realizes that the substance "which to the senses seems to be continuous in all its parts is in fact not so." The supposedly smooth surface is found to be lined with deep passages, which allow the ray to penetrate the surface and pass through the medium. Refraction, then, occurs when a light ray breaks through the supposedly smooth surface, dissolving the confusing paradoxes that guarded it secrets.

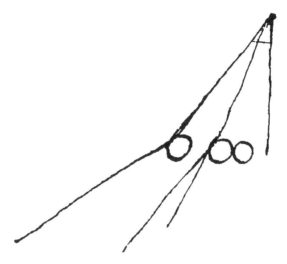

Figure 13.   Thomas Hariot, rays of light reflecting off atoms. British Library Add. MS 6789,
f. 181v (Courtesy of the British Library).

The guiding presence of the master narrative of exploration in Hariot's
reconstruction of refraction is unmistakable. The optical medium is perceived
much like Virginia or Guyana, as hiding great secrets beyond a supposedly
impenetrable barrier and confusing "mistes" that prevent one from "seeing"
clearly. The material medium appears at first glance to rebuff all attempts to
penetrate its mysteries, much like the barrier islands off the coast of Virginia
blocked the explorers' attempts to reach the wondrous interior. In both cases,
initial efforts to break through meet with a quick rebuff: the fate of hasty
interlopers who attempted to break through the Carolina Banks is marked
by their sunken ships, and the ray of light is reflected right back from the
material medium in its initial attempt at penetration. On further inspection,
however, the material medium, like the unexplored lands, is found to be permeated by deep passages, which allow access to its hidden mysteries. Light
can now pass through the medium, and Hariot is free to chart its inner recesses, just as he had mapped Virginia and Guyana. Hariot's maps of the newly
discovered lands and his reconstruction of the optical medium were both carried out in accordance with the master narrative of exploration. The inner
structure of matter thus acquired the same topography as unexplored lands
overseas.

Hariot made the guiding metaphor of his theory of refraction explicit when he issued a challenge to Kepler at the end of his letter: "I have now led you to the doors of nature's mansion where her secrets are hidden. If you cannot enter on account of their narrowness, then abstract yourself mathematically and contract yourself into an atom, and you will enter easily. And when you have come out tell me what wonders you have seen."[67] The passage of a ray of light through the material medium is here identified directly with a voyage of discovery into nature's hidden secrets. Kepler and the ray of light are both challenged to penetrate to the wonders that lie within. Both are at first rebuffed by the apparently continuous surface that bars the way to the interior, and both eventually locate the passages that allow them to penetrate to the innards of matter. Finally, on their exit, both must report everything they have seen to Hariot. Kepler's assigned role here is no different than Hariot's own role as Raleigh's spy in Virginia, charged with observing everything with "Argus eies." Like Hariot, Kepler must face the formidable obstacles guarding the hidden wonders, find passages to the interior and, on his return, report his findings to those who stayed behind. The search for the cause of refraction, like the larger hunt for the secrets of nature, are clearly no different than a geographical voyage of discovery.

The language that Hariot uses here indeed seems as if it were taken directly from the rhetoric of Raleigh's imperialist ventures. It merits comparison with a passage from Lawrence Keymis's *Second Voyage to Guiana*, discussed in Chapter 2.[68] Keymis, while decrying his countrymen's incredulity and enjoining them to invade Guyana, describes the invasion in terms similar to those that Hariot uses to describe the penetration of the material medium. The narratives are, indeed, almost identical: nature hides her treasures beyond the mountains of Guyana as well as behind the smooth material medium. From the outside the obscured secrets appear inaccessible: men would wonder at the strength and "marveilous" fortifications of Guyana, just as Kepler would be held up by the "narrowness" of the entrance. The persistent seekers of "truth" will, nonetheless, break through the fortifications, squeeze through the seemingly narrow passages, and win access to nature's treasures.[69]

In his poem prefacing *Achilles' Shield*, Chapman compared Hariot's search for knowledge to an exploring ray of light, breaking through "errors" and "mistes" and exposing the underlying harmony "to every comprehensive eye." Nathaniel Torporley described Hariot in 1602 in a similar manner, calling him a man "born to dissipate, by the splendor of undoubted truth, the philosophical clouds in which the world has been enveloped for many centuries."[70] In his theory of refraction Hariot turns this metaphor of discovery into an actual optical theory. The refracted ray acts just like the enterprising explorer or the investigating natural philosopher: it breaks through seemingly unsurpassable obstacles, penetrates to the interior, and (literally) brings to light what was previously obscured. For Hariot, the rhetoric of discovery was more than a general metaphor for the search for knowledge: his theory of refraction was the optical embodiment of an actual voyage of discovery.

"THE ENTIRE SCIENCE OF THE REFLEXION OF BODIES"

The Briefe and True Report on the New Found Land of Virginia was the only work Hariot published during his lifetime. Scattered among his papers and correspondence can be found repeated promises to publish his work, and some of his completed treatises were clearly intended for the printing press. Despite these good intentions, Hariot's scientific work remained entirely in manuscript form at the time of his death in 1621.[71] In his will Hariot left his papers to his friends Nathaniel Torporley, Thomas Aylesbury, and John Protheroe and instructed them to publish what they could.[72] The three, however, pursued their friend's dying request somewhat halfheartedly: the only tangible result of their efforts was the posthumous publication in 1631 of a short book based on his algebraic notes—the *Artis analyticae praxis*.[73] The book dealt with numerical solutions to algebraic equations and was based on a very small sample of Hariot's manuscripts. The vast majority of Hariot's manuscripts remain unpublished to this day.[74]

It would be unfair to judge Hariot's executors too harshly for their failure to follow up on his last wishes. As they exist today, the British Library manuscripts of Hariot's papers consist of thousands of folios, bound together in the semirandom order in which they were delivered to the British Museum by Lord Egremont in 1810. Most of the manuscripts are in the form of scattered notes in Hariot's nearly illegible hand, and only rarely does one encounter a batch that resembles a coherent tract. Furthermore, even the pages of these select treatises are often scattered in different locations and separate volumes of the collection. Hariot's papers in the Petworth House collection are somewhat more organized, having been carefully selected by Baron von Zach in 1786. Even here, however, in the absence of ordered treatises, it is usually very difficult to decipher Hariot's intentions and meaning.[75] Even granting that the manuscripts were, undoubtedly, better ordered in 1621 than they are today, it is clear that the simple task Hariot assigned to his executors was in truth a daunting undertaking.

A noted exception to this rule is a batch of papers entitled "De Reflexione Corporum Rotundorum Poristica Duo," which are found in the Petworth

House collection.[76] Almost unique among Hariot's papers, this batch forms a coherent (although far from complete) treatise on a single topic, namely, the collision of round bodies. The presence of this surprisingly finished text is explained by a letter from Hariot to his patron, the Earl of Northumberland, dated June 13, 1619, which is bound with a copy of the tract in the Harley manuscripts in the British Library:

> Sir,—When Mr. Warner and Mr. Hues were last as Sion, it happened that I was perfecting my auntient notes of the doctrine of reflections of bodies, unto whom I imparted the mysteries thereof, to the end to make your lordship acquainted with them as occasion served. And least that some particulars might be mistaken or forgotten, I thought best since to set them down in writing, whereby also now, at greater waight, you may thinke and consider of them if you please. It had been very convenient, I confess, to have written of this doctrine more at large, and particularly to have set downe the first principles, with such other of elementall propositions, as all doubtes might have been prevented; but my infirmitie is yet so troublesome that I am forced . . . to let alone till time of better abilitie. In the meane time I have made choyce of these propositions, in whose explication you shall find, the summe of all that of this argument is reasonable to be delivered.[77]

"De Reflexione," then, is a pedagogical text, a brief summary of older work intended to instruct the earl in the "mysteries" of reflecting bodies. It is characteristic of Hariot that even when consciously bent on writing a coherent treatise, he limited its scope to the barest minimum. His claim of ill health in 1619 may have been well founded, as the cancer, of which he was to die two years later, had already set in. One may, nonetheless, be somewhat skeptical of Hariot's claims, since he had used the same ploy in his correspondence with Kepler thirteen years earlier and for the same purpose—as an excuse for the brevity and incompleteness of his text.[78] Ill health, after all, never prevented Hariot from proceeding with his own work and compiling his imposing collection of unedited papers. It seems, rather, that he simply preferred to work on his own problems at his own pace, while leaving to his executors the imposing task of converting his notes to publishable form.

Hariot's version of the "mysteries" of reflecting bodies has been the subject of considerable scholarly attention in recent years.[79] Although the technical aspects of Hariot's theory are not my main concern here, a brief account would be helpful. Hariot's purpose was to predict the courses of two balls of unequal weight after they collide with each other at a given angle (see Figure 14). For the sake of the example, Hariot posited that the two balls' weights were at a 2:3 ratio to each other.[80] He started out by dividing the

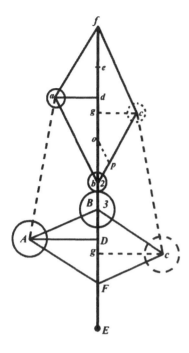

Figure 14.   Thomas Hariot, the collision of two unequal balls on a plane surface. The original is in the Petworth House collection, manuscript HMC 241/iv, f. 23 (Courtesy of the Earl of Egremont).

movement of the balls into two components: the first element, *db* and *DB,* respectively, is along the line defined by the centers of the two balls at the moment of collision. The second component, marked as *ad* and *AD,* is perpendicular to that line. Hariot assumed that the speed along *ad/AD* is fixed and unaffected by the collision. The impact is felt only along the line defined by the centers, marked as *fedgbBDGFE.*

The effect of the collision on the movement of the balls along this line is to be measured, according to Hariot, in terms of what he calls "nutus"—the tendency of a body to move in a certain direction.[81] Taking the smaller of the two balls, Hariot stated that the total nutus is composed of three motive acts. The first of these is the result of the smaller ball reflecting off the larger one. This provides the nutus *bd,* which is equal to *db,* the original component of the movement of the ball, but in the opposite direction. The second motive act results from the transfer of movement from the larger ball to the smaller one. The motion *BD* of the larger ball is added to the nutus *bd* and marked as nutus *de.* For the third motive act, Hariot considered the relative magnitudes of the balls. Since the excess of weight that the larger ball has over the smaller one is one third of its own weight, Hariot concluded that the combined nutus resulting from the first two motive acts, *be*, should be increased by one third of its length thus adding the segment *ef*.

The entire nutus of the smaller ball following the collision is therefore marked *bf*. When this is added to the original nutus in the perpendicular direction, *ad*, the resulting movement of the ball will be toward the point *c*. The same procedure is then applied in calculating the nutus of the larger ball, with the sole exception that rather than adding one third of the total in the final motive act, one third is subtracted. The total nutus of the larger ball is, then, *BF*, and the eventual direction of the ball is in the direction of the point *C*.

Both Jon Pepper and Martin Kalmar, who have studied Hariot's text in recent years, point out that Hariot's theory is flawed and inconsistent, particularly in his calculation of the third motive act.[82] It is more important for our purposes, however, to trace Hariot's reason for developing this theory, rather than his errors. Pepper, arguing that the medieval term "reflexion" was used almost exclusively in optical contexts, suggests that Hariot's concern with reflecting bodies is related to his theory of refraction.[83] This seems highly plausible, especially if we note that Hariot had written Kepler that "refraction is nothing else but internal reflexion."[84] It would be perfectly consistent for Hariot to try to develop a theory of reflexion, which could then be incorporated into his optical studies. The unmistakable similarity between the diagrams illustrating Hariot's theory of refraction and those illustrating the colliding balls gives further credence to this view. The optical sketches (see Figures 10–14) are rather sketchy, giving a qualitative impression of the rays passing through a medium, while the "colliding balls" diagrams attempt to be quantitatively precise (see Figure 14). Both, however, depict a crooked course resulting from collision with round bodies, and both do so by outlining the path of the reflecting bodies. It would hardly be a stretch to conclude that Hariot's kinematics were closely associated with his views on optics and were intended to complement them.

In particular, it would seem that Hariot's purpose in his theory of collisions was to provide a closer view of the interior of a material medium. After positing that light reflects back and forth within matter, he now sought to "see" this process up close and to account for its operation in detail. In essence, he was refining his former position by substituting a well-focused view of the happenings within matter for his previous generalized impressions. The attempt to map out and describe hidden realities is a typical Hariot move: he did the same when exploring and mapping Virginia and Guyana, as well as when he constructed his original refraction theory. The desire to actually "see" into the interior by bringing to light what was previously hidden, so characteristic of Hariot's work, here manifests itself in his theory of collisions.

Hariot's true intentions in his theory of reflecting bodies become abundantly clear in his introductory statement to the "De Reflexione":

> These porisms, by their power, delineate the entire science of the reflexion of bodies. And he who rightly comprehends them is a master, as it were, of all other cases and of the entire doctrine. And therefore not improperly they can be called masteries. They indeed rank among the principal elements which lead to the understanding of the innermost parts or mysteries of natural philosophy. Aristotle, and authors ancient and modern, have proposed problems of this kind. They ask the means, they argue, they conclude. But as the illustrious Terence said: 'Does not their understanding show that they have no understanding?'[85]

The guiding metaphor of Hariot's theory is explicit here and is reminiscent of his stance on his optical work or his geographical investigations. The standard narrative of exploration, in operation in Hariot's work on geographical discovery, is present here as well. Secret and valuable knowledge is hidden in "the innermost parts or mysteries of natural philosophy," hidden behind the impenetrable surface of matter. The way there is difficult and confusing, and many great authors, ancient and modern, have tried and failed to crack the secret of reflexion. The task seems hopeless until Hariot introduces his treatise. The "De Reflexione" is Hariot's "magisteria"—a guide to "all cases" and "the entire doctrine. It is a map of the way leading to knowledge and mastery of the hidden mysteries of reflexion.

In his letter outlining his theory of refraction, Hariot called on Kepler to contract into an atom and enter "nature's mansion where her secrets are hidden" so that he may return and tell what wonders he has seen.[86] The "De Reflexione" is precisely the kind of traveler's report Hariot was expecting in response. It provides an unmediated close-up view of the happenings within the hidden interior of matter. By putting it into writing and diagrams Hariot had produced a guide to this hidden realm in much the same manner as he had produced his charts of the undiscovered lands of Virginia and Guyana. The rhetoric is similar and the underlying discovery narrative of exploration is similar. Even the end result—the division of the smooth coastline or material medium into discrete components separated by open passages is similar. The tale of exploration and discovery, which had structured Hariot's geographical and optical work, is here guiding his vision of the inner workings of matter.

CHAPTER 5

"The Club of Hercules":
Hariot's Mathematical Atomism

A MATHEMATICS OF VISION

Hariot's involvement in the Elizabethan exploration enterprise left deep marks on his scientific thought—so much so that in much of his scientific work Hariot preferred a geographical explorer's perspective to more traditional approaches to natural philosophy. His emphasis on unmediated "seeing" as the true source of knowledge was characteristic of an explorer's point of view. After all, a voyager's truth claim was based uniquely on his firsthand experiences in distant lands.[1] Similarly, according to Hariot, "the truth when it is seen is known without other evidence," a creed that seems better suited to the exploration of foreign lands than to systematic investigation of nature. Despite the difficulties, however, Hariot remained true to his word. His work on optics and kinematics is structured around the notion that in order to understand the workings of nature one must actually "see" it in operation.

   Not only did Hariot insist on direct observation of the secrets of nature, but he was also quite confident that he knew what he would find. The standard tale of exploration, which guided the enterprising travelers on their voyages of discovery did much the same for Hariot in his scientific explorations. The passage of the ray of light through the material medium was for Hariot akin to a geographical expedition, bent on penetrating all and leaving nothing uncovered. Like the true explorer, the ray was at first rebuffed by the confusing array of defenses protecting the hidden secrets; and like the explorer, it eventually found an open passage leading through the medium and revealed its wonders. In the process, matter acquired a structure that was strikingly similar to the new geography of the foreign lands. Both appeared superficially inaccessible, impenetrable, and defended by insurmountable obstacles; and both were, in fact, traversed by many open passages. The maritime culture of exploration provided Hariot with both an epistemological approach and a guiding narrative for conducting his investigations. In doing so, it helped shape the theoretical content of his work as well.

But what of mathematics? Although a reliance on sight is a problematic methodology for the study of nature in general, it is a particularly improbable approach to the study of mathematics. Mathematical objects, as traditionally construed, can never be seen. For Platonists, they belong to the perfect, transcendent realm of ideas. For orthodox Aristotelians they are nothing but generalizations based on sense experience. In either case they are not sensible objects, but rather mentally constructed abstractions. The knowledge we have of objects such as geometrical figures comes not from "seeing" them—which is impossible—but from rigorous logical reasoning about their nature and properties.

Hariot's emphasis on the importance or "seeing" for knowing the truth stands in direct opposition to this commonly accepted conception of mathematical knowledge. In the traditional view, the unique power of mathematical demonstrations lies in the fact that they are *not* based on sense perception, but rather on exact reasoning from first principles.[2] For a mathematician to privilege sight over "other evidence" would appear nonsensical. The truth of mathematics is guaranteed precisely by its sole reliance on logical reasoning at the expense of the unreliable and misleading senses.[3] It is hardly surprising that geographical explorers would emphasize the importance of personal observation. It is far more perplexing to find the same view expressed by a celebrated mathematician.

Hariot, nevertheless, remained true to his word in his mathematical activities, just as he was in his optics and mechanics. The desire to "see" is as much in evidence in his discussion of the equiangular spiral and the structure of the continuum, as it was in his geographical accounts of America and his studies of matter. Mathematical constructs, for Hariot, were no different than any other objects, and he meticulously dissolved them into their component parts in order to examine their structure. In Hariot's view, true understanding of mathematical entities, just like physical ones, would follow from the close inspection of their inner structure.

Furthermore, in Hariot's mathematics, just as much as in his physical investigations, the structure thus revealed is by no means random. It is, rather, shaped by the standard narrative of geographical exploration. The confusing array of difficulties to be overcome, the hidden mysteries of the interior, and the eventual detection of clear passages through the obstacles are clearly in evidence in Hariot's mathematical studies. The structure of the mathematical continuum, much like the optical medium, thus came to resemble the geography of foreign lands.

The structural similarities between Hariot's mathematical, physical, and geographical investigations seem less surprising if we note that Hariot did not strictly distinguish between mathematical and physical objects. Since Hariot left us no structured tracts on the topic, it is difficult to trace the exact details of his position on the relation between the mathematical and the physical. In reconstructing his views, let us first go back to his challenge to Kepler to penetrate the hidden secrets of nature: "I have now led you to the doors of nature's mansion where her secrets are hidden. If you cannot enter on account of their narrowness, then *abstract yourself mathematically and contract yourself into an atom*, and you will enter easily."[4] Hariot's words are revealing. The voyage into the secrets of nature is essentially a physical one: Kepler must turn himself into an atom, enter the optical medium, and explore the mysteries hidden within. The manner to achieve this transformation, however, is to "abstract yourself mathematically." Mathematical knowledge, as understood here, does not deal with disembodied abstractions; it is closely interwoven with the material world. In fact, Hariot implies, mathematical truths are inseparable from physical reality. By abstracting himself mathematically, Kepler would achieve an inside view of the structure of the material medium, as if he were an atom traveling through it. A mental voyage of mathematical abstraction, it seems, is no different than the physical voyage of an atom within matter. Mathematics thus replicates the structure of physical matter, and knowledge of one implies knowledge of the other.

The challenge to Kepler appeared in the context of Hariot's physical investigations—in this case his atomistic theory of matter, which he used to explain the optical phenomenon of refraction. It is significant to note that the same view of the relationship between matter and mathematics also appeared in the context of Hariot's mathematical studies. In his unpublished tract "De Infinitis," Hariot argues for the existence of a maximum limit of multiplication and minimum limit of divisibility of mathematical magnitudes: "That such quantity which I call universally infinite, hath not only act rationall, by supposition, or by consequence from mere supposition: but also act reall or existence: in an instant, being a perfect actual being, or in time, passed by motion both finite and infinite: with many reall consequences or properties consequent."[5] Mathematical magnitudes, for Hariot, are clearly not abstract, hypothetical structures. They "act reall or existence" and have "many reall consequences or properties." Just as in the letter to Kepler on optics, Hariot is here committed to a direct correspondence between mathematical truths and the sensible world. The analysis of mathematical magnitudes produces true knowledge about physical reality.[6]

Hariot's statements on the correlation between mathematics and the physical world fall well short of a clearly delineated philosophical position. Many questions remain unanswered. What is the source of this correlation? Is mathematics modeled on the physical world, or is physical structure necessarily ingrained with mathematical truths? There is no way of knowing Hariot's views on these fundamental questions. Nonetheless, once it is acknowledged that for Hariot mathematical truths and physical reality were almost interchangeable, it is far less surprising to find that his views on mathematical magnitudes closely resembled his theory of the structure of matter. In particular, Hariot's reconstruction of the mathematical continuum was unmistakably similar to his views on the surface of an optical medium.

Although Hariot's position on the relationship between mathematics and the physical world points to a close similarity between the two, this does not mean that he arrived at his mathematical positions through simple analogy with his views on matter. Hariot's work on the mathematical continuum, in fact, has its own unique genealogy and can be traced in considerable detail. He arrived, for example, at his mathematical atomism in the context of his cartographic work on the true chart rather than as a result of his optical or material investigations. It is clear, nevertheless, that Hariot viewed physical and mathematical problems in similar terms, approached them with a similar methodology, and resolved them in a similar manner. In mathematics, just as in his optics and mechanics, Hariot viewed his investigations as voyages of discovery in search of great hidden treasures; in both cases he resolved his difficulties by breaking down the smooth continuous surface and positing deep passages that pass right through it. The master narrative of exploration, which guided Hariot and his accomplices in their exploration of foreign lands and had proved equally effective in revealing the secrets of the optical medium, is here applied to the study of "pure" mathematics.

The focus here is on Hariot's treatment of one of the classical problems of mathematics—the structure of the continuum. The knotty problems and paradoxes of the continuum have been known since antiquity and they were to be of crucial importance in the development of the infinitesimal methods leading to the calculus. Hariot's interest in these matters arose from his work on the Mercator projection for navigational charts. While calculating the tables necessary for the construction of such a chart, he constructed an equiangular spiral, composed of an infinite number of sections bearing a fixed ratio to each other. The difficulties he encountered in this construction led him to a detailed study of the fundamental problem of the mathematical continuum.

In his most systematic treatment of the continuum, in the unpublished tract "De Infinitis," Hariot treated the problem in accordance with the standard narrative of exploration: great mysteries were perceived to be hidden beyond the apparently impenetrable surface of the continuum. They were protected by a confusing labyrinth of paradoxes and appeared unattainable at first glance. Despite the difficulties, Hariot insisted that the seemingly smooth surface is in fact composed of indivisible sections with breaks in between. These allow Hariot to break through the confusion and master the secrets of the continuum.

Hariot arrived at the problem of the continuum through his involvement in the exploration project. The construction of a chart that would preserve true directions across the globe was a major practical problem for the transoceanic voyages of the age, and Hariot applied himself to it with great energy. When he encountered difficulties, Hariot, like many other contemporary mathematicians, adopted the guiding narrative of his enterprising patrons and chose to view mathematics itself as a voyage of discovery.[7] The mathematical continuum became an undiscovered country in its own right, and Hariot approached it with the guidance of the standard tale of exploration. It is hardly surprising that, in the process, the mathematical continuum came to resemble the geography of lands overseas.

THE CONTINUUM

The structure of the geometrical continuum was a central problem of Greek mathematics since at least the sixth century BC. The Pythagoreans, who adhered to the maxim that "all is number," at first considered integers (or collections of units) to be the fundamental building blocks of the cosmos. Accordingly, any continuous magnitude was considered to be a conglomeration of its fundamental component parts and could be assigned an integral number accordingly. It was a basic tenet of this position that since all magnitudes could be dissolved into those fundamental building blocks, all magnitudes are, in principle, commensurable.[8] This early Pythagorean view floundered with the discovery that the side of a square and its diagonal are incommensurable. In other words, no matter how many times the side of a square and its diagonal are divided, they will never reach a component magnitude that they both share.[9]

The discovery of incommensurability was a serious setback to the atomistic conception of geometrical figures. Lines could not be conceived of as composed of points, since then such points would serve as a common component magnitude that all lines shared. Similarly, planes could not be construed as being composed of lines, nor volumes as composed of surfaces. Mathematical atomism, so appealing to the materialistic intuition of geometry, seemed to have suffered a mortal blow.

The problem of incommensurability was not the only difficulty encountered by the Greeks in their attempts to understand the nature of the continuum. Since continuous magnitudes were not composed of an integral number of atomistic components, it was concluded that the continuum must be infinitely divisible. Even this position, however, came under attack in the four famous paradoxes attributed to Zeno the Eleatic.[10]

Two of those paradoxes, known as the Dichotomy and Achilles, concern us here. In the Dichotomy, Zeno argues that before a given distance can be traversed, half of it must be traversed first. Before this—half of half the distance, and so on to infinity. The implication is that an infinite number of segments must be crossed before any given distance is traversed. Since this is impossible in a finite amount of time, no movement is possible. The Achilles paradox posits a race between swift Achilles and the slow tortoise and follows a similar line of reasoning. The tortoise begins the race a certain distance ahead of Achilles. By the time Achilles reaches the tortoise's original position, the latter has already moved on. When Achilles reaches the new position the tortoise has already moved yet again, and so on to infinity. The conclusion was that regardless of their relative speeds, Achilles would never catch up with the tortoise, since he would have to traverse an infinite number of segments to do so. Since the paradoxical results in both the Dichotomy and Achilles were reached under the assumption that the continuum was infinitely divisible, it must be concluded that this assumption is false.

The investigation of the nature of the relationship between the continuum and its parts thus ended with what appeared to be an inextricable paradox: the continuum was not composed of a definite number of parts, since that contradicted the established fact of incommensurability. Conversely, it could not be infinitely divisible, since Zeno had shown that such an assumption leads to absurd results. Aristotle, who defended infinite divisibility, countered by claiming that Zeno was confusing infinity by division, and infinity by composition. This seemed to imply that while a continuous magnitude could potentially be infinitely divided, it should not be assumed that it could be recomposed from an infinite number of real parts.[11] Practicing mathematicians followed Aristotle's lead and attempted to defuse the problem. Eudoxus of Cnidus (ca. 370 BC) and Archimedes of Syracuse (d. 212 BC) in particular, developed highly effective mathematical techniques such as the "method of exhaustion" that consciously avoided treating continuous magnitudes as composed of discrete parts.[12] The problem of the continuum, it was shown, could indeed be circumvented. It could not, however, be solved.[13]

The question of the composition of the continuum reappeared at the forefront of mathematical thought during Hariot's lifetime and remained there throughout the seventeenth century. This was due to a renewed interest in Archimedean problems, such as the area enclosed under conic sections, the quadrature of various geometrical figures, and the tangents of curves. In treating such problems, Archimedes himself was not averse to the use of infinitesimals: the length of a curve could be calculated by treating it as a con-glomeration of points; the area of a surface was arrived at by treating it as the sum of its "component" lines; and tangents could be found by estab-lishing the direction of a curve at one of its "component" points.[14] Clearly, the assumptions made in these calculations completely ignore the problems of the continuum, and the paradoxes that result from treating it as a con-glomeration of discrete parts. Archimedes, therefore, while accepting these methods as reliable for practical purposes, also insisted that they were inher-ently flawed and could never be used for proper mathematical demonstra-tions. For true demonstrations, Archimedes resorted to the geometrically rig-orous method of exhaustion.[15]

Early modern mathematicians proved far less scrupulous in their use of infinitesimals than their ancient predecessors. Simon Stevin of Bruges (1548–1620) was among the first to introduce infinitesimal considerations in his *Elements of the Art of Weighing* of 1586.[16] In proving that the center of grav-ity of a triangle lies along its median, Stevin summed up the parallel lines composing the sections of the triangle on both sides of the median and showed them to be equal. Rather than resort to the geometrically rigorous method of exhaustion, Stevin was quite satisfied to treat the surface of the triangle as if it were composed of lines.[17] Kepler developed a similar approach in his *Nova Stereometria* of 1615, where, for example, he calculated the area of a circle by treating it as a polygon with an infinite number of sides.[18] Galileo devoted considerable attention to the problem of the continuum in his *Two New Sciences*, where he appeared to base his model of the geomet-rical continuum on an atomistic intuition of matter.[19]

Galileo's speculations were developed into a full-fledged mathematical method by his student, Bonaventura Cavalieri (1598–1647). In his *Geo-metria indivisibilibus* of 1635 and the *Exercitationes geometricae sex* of 1647, Cavalieri addressed a vast array of new problems, dealing mostly with areas and volumes.[20] In doing so, Cavalieri was ignoring the well-established paradoxes of the continuum and the accepted canons of mathematical rigor. "It is manifest," he boldly announced, "that plane figures should be con-ceived by us like pieces of cloth are made of parallel threads. And solids are like books, which are composed of parallel pages."[21] Continuous magni-tudes, in other words, were composed of fundamental indivisible parts after all. This challenge to the mathematical tradition was quickly answered by

Paul Guldin in his *Centrobaryca*, who charged that Cavalieri was undermining the foundations of mathematical knowledge.[22] In response, Cavalieri changed the format of his method somewhat, but never retracted his fundamental position. For Cavalieri, the continuum was in truth an aggregate of its component indivisible parts.

The "method of indivisibles," as Cavalieri named the infinitesimal approach, dominated seventeenth-century mathematics. Evangelista Torricelli (1608–47) continued to develop Cavalieri's work in Italy, while Pierre de Fermat (1601–65), Blaise Pascal (1623–62), and Gilles Personne de Roberval (1602–75) addressed similar issues in France, to name but a few major figures. John Wallis (1616–1703) and Isaac Barrow (1630–77) revived the mathematical tradition in England in the 1660s by bringing their own perspective to bear on continental mathematics: Wallis emphasized a loose, nonrigorous algebraic approach to infinitesimal calculations, while Barrow insisted on the primacy of well-defined geometrical representations.

The crucial importance of the method of indivisibles in the history of mathematics lies, of course, in the fact that it was eventually transformed by Newton and Leibniz into the modern calculus. By developing their algorithms, Newton and Leibniz abstracted the method of indivisibles from its dependence on any particular geometrical problem. The "method," as it was used throughout most of the seventeenth century, was essentially a way of solving problems of areas, volumes, tangents, and so forth. It provided a suitable and fruitful approach for handling problems that were essentially geometrical. This, however, changed with the introduction of the calculus: rather than treat separate geometrical problems, Newton and Leibniz focused on the internal coherence of the method itself.

The inverse relationship between problems of area and volume, on the one hand, and problems of tangents, on the other, was known to Isaac Barrow and probably to others as well. With Newton and Leibniz, this relationship was detached from its geometrical roots and became the algebraic inverse relationship between differentiation and integration. This relationship was the fundamental insight of Newton's and Leibniz's calculus. Any function (within defined limits), according to Newton, had a "fluxion" and was itself a "fluxion" of a different function. These relationships could be applied to various kinds of problems, including geometrical ones. Unlike the old method of indivisibles, however, they were not dependent on geometrical representation: the calculus was an abstract, universally true set of mathematical relations, which required no geometrical interpretation. With the development of the calculus, the former method of indivisibles was transformed into an abstract algorithm, which could be used as a flexible and reliable tool in any area of investigation. Stevin and Cavalieri's bold attack on the paradoxes of the continuum had culminated in the founding achievement of modern mathematics.

MERIDIONAL PARTS AND THE ADDITION OF SECANTS

The renewed interest in the problem of the continuum in early modern mathematics should be kept in mind when reviewing Hariot's work. When viewed against this background, Hariot's achievement seems impressive indeed. A full generation before Cavalieri, Hariot was busy analyzing the properties of a unique curve—the equiangular spiral.[23] In calculating the length of this curve, the area it encloses, and the sum of its radius vectors, Hariot employed the kind of approach that would become the hallmark of the method of indivisibles. He approximated the continuous spiral by an infinite succession of straight segments. In Hariot's construction, the smooth continuous spiral was dissolved into its discrete component parts, in clear defiance of the well-established paradoxes of the continuum.[24]

It is characteristic of Hariot's circumstances and interests that he arrived at the consideration of the spiral through his involvement in maritime affairs. The production of a naval chart that would preserve true directions became an urgent concern with the rapid increase in transoceanic travel in the sixteenth century. The traditional plane chart was quite sufficient for navigation along the coasts of Europe. Its depiction of equally spaced parallels and meridians was a good approximation of conditions in southern Europe and the Mediterranean. In northern Europe, the relatively short distances made the error of the chart easily correctable by assuming a fixed ratio between the length of a degree of latitude and a degree of longitude.

In transoceanic voyages, however, the errors were cumulative and serious. As discussed above in Chapter 3, a navigator who would attempt to calculate his landfall after a transoceanic voyage at a fixed compass bearing by using the plane chart would find himself to be in grave error. His north-south progress would be correctly estimated, since the latitudes are equally spaced on the actual globe as well as on the plane chart. Not so with regard to the east-west departure: the chart assumes that the meridians are equally spaced at distances equal to those of the parallels. In reality, a degree of longitude is only equal to a degree of latitude at the equator, and it gradually contracts as the meridians converge on the poles. As a result, our flat earth navigator would find that he had seriously underestimated his east-west progress. Since the length of a degree of longitude shortens as one travels further north (or south) from the equator, he would, in fact, be traversing many more degrees of longitude than his chart would allow. How, then, could a chart be constructed in which a ship's course at a fixed bearing would be truly represented by a straight line?[25]

The solution to this problem was already recognized by Gerard Mercator in his world map of 1569. Rather than attempt to imitate the true globe by depicting the meridians as converging, Mercator chose to leave them as they were on the plane chart—as equally placed parallel lines. In order to compensate for this error, Mercator gradually increased the distance between the lines of latitude. At the equator a degree of latitude was equal to a degree of longitude, but as one progressed toward the poles, the length of each degree of latitude was increased. As a result, the ratio between a degree of latitude and a degree of longitude at any given point on the chart was equal to the corresponding ratio on the globe. A straight line at a given angle drawn on the chart would, therefore, cross the same number of longitude degrees for every degree of latitude traversed as would the true oceangoing mariner, sailing on the corresponding fixed compass bearing. True directions, in other words, would be preserved.[26]

The main problem with Mercator's method was one of calculation: By how much should a parallel degree be increased at a given latitude in order to preserve true directions? Mercator himself never explained his method, leaving others to develop their own approaches. John Dee constructed tables that suggest that he was working on the problem as well, but he gave no details of his calculations either.[27] The first mathematician to publish detailed tables (later known as "meridional parts" tables) for the construction of Mercator projection maps was Edward Wright, in his *Certaine Errors in Navigation* of 1599.[28] Even more importantly, the main part of his book was dedicated to a detailed explanation of his method for calculating the necessary figures.

Wright's method was essentially simple and was probably similar to the ones used by Mercator and Dee. It involved a process that came to be known as "the addition of secants."[29] The secant of a given degree of latitude gives the ratio between the radius of the globe as a whole, and the radius of the parallel at that particular latitude. Since the radius of the globe is equal to the radius of the equator, and since the circumference of a circle is proportionate to its radius, the secant provides the ratio of the circumference of the equator to the circumference of the latitude line at the given degree. A degree of longitude at a given latitude is the 360th part of the circumference of that parallel. The secant is, therefore, the ratio of a longitude degree at the equator to a similar degree at the given latitude. Furthermore, because a degree of longitude at the equator is equal in magnitude to the fixed degree of lati-

tude, the secant of a given latitude is also the ratio between a degree of latitude and a degree of longitude at that given latitude on the globe. This, of
course, is precisely the ratio by which a latitude degree must be "stretched"
in order both to maintain the true ratio between latitude and longitude
degrees at that point on the globe and to preserve true directions.

In order to calculate by how much the entire map should be "stretched" in
the north-south axis in the Mercator projection, one must simply add the secants to one another at regular intervals (see Figure 15). At the equator, where
the degrees of latitude and longitude are exactly equal (that is, sec0° = 1), the
latitude degrees will not be increased at all. But with each successive secant
being added for 1°, 2°, 3°, and so on, the degrees of latitude gradually stretch
out. The 10° N. (S.) parallel will be displaced 3' north (south) of its position
in a plane chart; the 20° parallel will be displaced by 25'; and the 30° parallel will be displaced by 1.5°. By the time the higher latitudes are reached,
the cumulative displacement becomes far more significant: the 60° parallel
in Wright's projection will be marked where the 76° line would be in a plane
chart; the 70° line would be marked at 99°; and 80° would be marked at
140°. These last two, of course, would far exceed the confines of any plane
chart, which would be limited to 90°.[30]

The accuracy of Wright's method was determined by the size of the regular intervals of latitude at which he calculated the secants and added them up.
The smaller the interval, the greater the accuracy of the projection. In the
table in *Certaine Errors*, Wright added the secants at 10' minute intervals.
This was perfectly sufficient for practical cartographic purposes. The accuracy of a chart produced with Wright's figures would be well within the
bounds of precision allowed by the measurement technology of the age.

In reading Wright's account of his project, it becomes clear that he considered the general aims of a mathematical practitioner to be fundamentally
similar to those of a geographical explorer. Both attempted to break through
obstacles and open the way to great riches. But what of the actual method
suggested by Wright: Can the technical details of his method be shaped by the
rhetoric of exploration? We have already seen how Hariot's work on optics
and dynamics was informed by the standard narrative of exploration. Can the
same be said of Wright's method of the "addition of secants"?

In examining this issue, it should be noted that the method that Wright
hails as the breakthrough in the study of labyrinthine plane chart consisted,
in his words, of having the map "easily and truly divided into parts." As we
have already seen, the geographical discourse of discovery entailed precisely this procedure. When the standard tale of exploration was applied to the
Canadian arctic by Frobisher, to Virginia by White and Hariot, or to Guiana

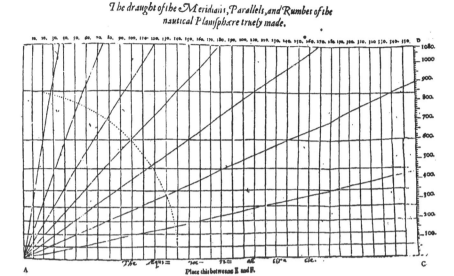

Figure 15.   Edward Wright's diagram of the Mercator projection. Printed in Edward Wright, *Certaine Errors in Navigation* (London, 1599).

by Raleigh and Keymis, the result was always the same: the smooth impenetrable coastlines were carved by rivers, straits, and narrows. In other words, they were indeed "divided into parts."

Was Wright's method, then, in any way shaped by the rhetoric of exploration? There is, undoubtedly, some evidence to that effect. Wright operated in social circles in which this discourse was extremely prevalent and even wrote a travel account himself.[31] He clearly defined his mathematical activities in terms of the narrative of discovery and viewed himself as an enterprising explorer who managed to break through the confusion of the plane chart. Finally, his procedure divided the nautical chart "into parts" in much the same way as the tale of exploration carved up newly discovered territories.

Nevertheless, although the combination of all these circumstances is indeed suggestive, it must be conceded that the cumulative evidence is less than conclusive. The intriguing analogies between the rhetoric of exploration as applied to mathematics and the mathematical procedure of the addition of secants in the end remain just that—suggestive analogies. We do not know enough about how Wright approached his project, or how he arrived at his solution, to definitely determine the exact interplay between his nautical rhetoric and his mathematical techniques. The case, however, is very different when we move from the published writings of Edward Wright to the unpublished works of his contemporary, Thomas Hariot. The volumes of manuscripts that Hariot left behind offer us a unique glimpse into the speculations that guided his technical mathematical approach.

THE EQUIANGULAR SPIRAL

Hariot's approach to the Mercator projection differed greatly from the route taken by his contemporaries. Edward Wright and John Dee focused their investigations on the particular cartographic problem at hand: How does one calculate the proper distances between latitude lines in order to create a chart that would preserve true directions? As has been noted, their (probable) common solution, the addition of secants, was quite satisfactory for mapmaking purposes. Furthermore, if more accuracy was required, it was only necessary to shorten the distances at which the secants were calculated and added up. The method was simple, allowed for easy calculation, and could be adjusted to any degree of accuracy a cartographer would require. It was a good solution to a practical problem.

Hariot was undoubtedly familiar with Wright's approach. It has even been suggested that in the early 1590s, long before Wright published his *Certaine Errors*, he constructed his own meridional parts tables by adding up secants.[32] Evidently, however, the method that was considered quite satisfactory by Mercator and Dee, and which Wright claimed broke through the labyrinthine errors of the plane chart, fell short of Hariot's exacting standards. Between 1594 and 1614 he devoted an enormous amount of work to the calculation of meridional-parts tables using a radically different approach.

Although Hariot nowhere explains the reasons for his dissatisfaction with Wright's method, the nature of his alternative approach is revealing. Hariot did not settle for a superficial solution to the practical problem. Instead, he sought to understand the fundamental inner structure of the problem at hand. The course sailed by a ship on a fixed compass bearing (or "rhumb"), he discovered, is produced by a succession of radius vectors centered on the pole, which bear a fixed ratio to each other.[33] This insight enabled him to analyze the rhumb in great detail and calculate its length, the area it encompasses, and other characteristics. Hariot's insistence on understanding the inner structure of the curve, however, did not stop at this level: the various relations between the radius vectors and the segments of the rhumb led him to look even further into the inner structure of lines in general. What started out as a practical problem of cartography, culminated in an investigation of the very nature of the geometrical continuum.[34]

Hariot, to put it simply, wanted to look inside. Wright offered a straight-forward external solution to a practical problem and proved its validity through Euclidean deduction. Hariot wanted to see the hidden inner structure of the curve, understand the source of the problem, and deduce a solution from there. When this proved to be difficult and elusive, he looked even more closely at the structure of continuous lines, attempting to discover the inner relations that govern their behavior. For Hariot, true knowledge followed from "seeing." His optical and mechanical studies attest to the profound effect the desire to "see" hidden inner structures had on the content of his theory of matter. The same is true here: Hariot was attempting to observe firsthand the inner structure of a rhumb line and the hidden secrets of the continuum itself. His views on the nature of geometrical objects were shaped through this very process.

Hariot's procedure here follows the same pattern as his investigation of optical refraction. In studying the optical medium, as he described it to Kepler, Hariot embarked on a voyage of discovery into the inner recesses of matter. In his investigation of the rhumb, Hariot ventured on a similar expedition—this time into the secrets of the equiangular spiral and the mysteries of the mathematical continuum. In his voyage into matter, Hariot observed discrete atoms and passages between them. In his mathematical venture, he discovered a similar landscape.

Hariot, appropriately enough, began his study of the Mercator projection with a hypothetical voyage of a ship, sailing at a constant bearing on the globe. The course of such a ship was obviously important for the navigators of the age, dependent as they were on the compass to determine their progress. Nothing could be simpler for a navigator at sea than to steer at a fixed compass bearing. Unfortunately, while the rhumb was easy to steer, it was also very difficult to track on a plane chart. The path it traced on the globe was a complex spiral, converging on the pole in an infinite number of cycles, but never quite touching it (see Figure 16). Gemma Frisius (1508–55) described the resulting course as follows: "The voyages by sea . . . are crooked, although guided by the Magnet, and are neither like to great circles nor yet to Parrallels: but only a kind of crooked lines, all of them at length concurring in one of the poles."[35] Hariot's contemporary, John Davis, "the Navigator," in his popular nautical manual *The Seamans Secrets* of 1595, refers to this type of sailing as "paradoxall navigation" and explains, "Paradoxall Navigation demonstrateth the true motion of the Ship upon any corse assigned . . . by which motion lines are described neyther circular nor straight, but concured or winding lines, and are therefore called paradoxall, because it is beyond opinion that such lines would be described by plaine

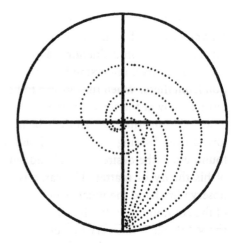

Figure 16. Nautical rhumbs of seven different bearings. Based on a diagram by Hariot in British Library Add. MS 6788, f. 478 (Courtesy of the British Library).

horizontall motion."[36] Clearly, the rhumb was considered paradoxical long before Hariot began studying it. The purpose of the Mercator projection was to construct a chart in which this confusing curve would be transformed into a simple straight line.

Hariot approached the problem of the paradoxical rhumb methodically.[37] In reality, the navigational rhumb is a complex, three-dimensional curve traced on the surface of the Earth's sphere. Analyzing such a curve directly with the mathematical methods of the time was practically impossible, as Hariot acknowledged. He therefore proceeded to reduce it to a more manageable curve—the equiangular spiral drawn on a flat surface. Using traditional Euclidean geometry, he proved that the equiangular spiral retains all the essential properties of the more complex rhumb. Any results he would obtain by analyzing the spiral would therefore hold for the true navigational rhumb as well.

The equiangular spiral is a curve that revolves endlessly around a central point, approaching it ever more closely but never actually reaching it. Its defining characteristic is that if straight lines were drawn emanating from the central point, the spiral would cross each and every one of them at the same angle. In this respect it preserves the defining characteristic of the navigational rhumb, which crosses all the lines of longitude emanating from the Earth's poles at the same angle (see Figure 17).

Hariot now began to analyze the characteristics of this curve. A true equiangular spiral, he knew, is a smooth, continuous, and rounded curve, with no straight sections or sharp angles. Nevertheless, in order to analyze it, he decided to treat it *as if* it were composed of straight lines at a fixed angle

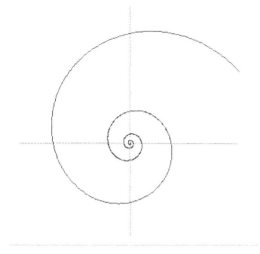

Figure 17. A modern depiction of an equiangular spiral.

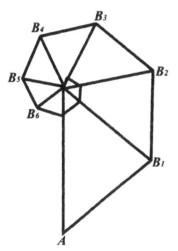

Figure 18. An equiangular "spiral" composed of similar triangles. Based on a sketch by Hariot in British Library Add. MS 6784, f. 248 (Courtesy of the British Library).

to each other. The resulting figure looked more like an ever shrinking polygon than a spiral (see Figure 18), but Hariot showed that making such an approximation had distinct advantages. The straight lines forming the sides of the approximated spiral, he pointed out, could be construed as the sides of triangles placed side by side, all of them with one of their angles anchored at the central point of the spiral. Most importantly, he showed that all those triangles were similar to each other, meaning that they all had the same three angles, and that their sides were at a fixed ratio to each other. More simply, one could say that all these triangles were of the exact same shape, except that they were of different sizes. The first one was the largest, whose side formed the most external part of the "spiral," the next one was adjacent to it and somewhat smaller, and the closer one got the central point, the smaller the triangles became.

Not only were the different triangles composing the approximate spiral similar to each other, Hariot showed, but their dimensions are also at a fixed ratio to each other. This was a crucial point. Let us assume, for example, that each side of the second largest triangle is half as long as each side of the largest triangle. Since this ratio is fixed, this means that the sides of the third largest triangle are in turn half as long as those of the second largest triangle, and those of the fourth largest triangle half as long as the sides of the third largest, and so on.

If we assume that a given side of the largest triangle is of length "1," then the side of the second triangle will be of length $1/2$, the side of the third triangle will be of length $1/4$, and so on. The sides of the triangles, in other words, can be described as

1, $1/2$, $1/4$, $1/8$, $1/16$ . . .

In mathematical terms one can say that the sides of the triangles form a geometrical series with a fixed ratio (or "β," in Hariot's terms) of $1/2$. Most significantly, the same figure that marks the fixed ratio between the external sides of the "spiral" also marks the fixed ratio between the lines connecting it to the central point, the radius vectors. This is because the radius vectors, like the external sides of the "spiral," are sides of the very same series of similar triangles and therefore are in the same ratio to each other.

In practice, Hariot found, the ratios don't work out quite so neatly as in our simple example. In order to calculate a true ratio, he constructed an approximate spiral that intersects with the radius vectors at the fixed angle of 45°. He calculated that if the radius vectors were drawn at 1' minute intervals (1' minute being the $1/60$th part of 1 degree), then the fixed ratio between the sides of the triangles would be an unwieldy $\beta=0.9970915409725778$. Undaunted, Hariot proceeded to calculate all the essential characteristics of the approximate spiral using this figure.

The feature of the spiral that most interested Hariot was the relative lengths of its radius vectors—the straight lines connecting the "spiral" to its central point. Since the radius vectors were at a fixed ratio of "β" to each other, then if one assumes that the first one, r_0, is equal to 1, then the next one, after a turn of 1', is equal to β, the second one, after a turn of 2' equals β^2, and so on. In other words, the radius vectors form a geometrical series in which

$$r_1=\beta, \ r_2=\beta^2, \ r_3=\beta^3 \ldots r_i=\beta^i \ldots$$

Most notably, the "i" in this series represents the number of minutes turned by the spiral to that point (1' for the first radius vector, 2' for the second, and so on). Hariot then proceeded with the enormous task of calculating tables that would give the value of β^i for every "i" interval.

Hariot had good reasons to focus on the series of radius vectors: by using traditional Euclidean means, he could calculate the correct radius vector "L" on the equiangular spiral that corresponded to any given latitude on the original, three-dimensional, nautical rhumb. For example, if Hariot was concerned with a point of 60° north on the nautical rhumb, he could calculate its "L," or the exact length of the radius vector that would correspond to 60° north on the two-dimensional equiangular spiral. In other words, any degree of latitude ϕ on the true nautical rhumb has a corresponding radius vector "L" in the equiangular spiral, which should fit in somewhere within the series of radius vectors. In mathematical terms each latitude ϕ of the nautical rhumb has an "L" such that

$$\beta^{i+1} < L < \beta^i.$$

Using his tables for "β^i" Hariot could now find the correct "i" for which this would hold.

Let us pause for a moment and consider the significance of this "i": it is the number of minutes that the spiral has turned before it reached the point where its radius vector was equal to "L." For the 45° nautical rhumb it is the number of lines of longitude (at 1' intervals) that the rhumb has crossed before reaching latitude ϕ.

We have already seen that the main difficulty with constructing a Mercator projection was determining how many degrees of longitude had been crossed for every degree of latitude traversed while sailing at a fixed bearing. Hariot has here solved this fundamental mathematical problem. For any degree of latitude, the figure "i" provides an excellent estimate for the number of degrees of longitude traversed up to that point.

Now, it will be recalled that the purpose of the Mercator projection was to preserve true directions by representing rhumbs as straight lines. Since in the Mercator chart the meridians are depicted as equally spaced parallels, it would be necessary to place latitude ϕ in the place of latitude "i" if the 45° rhumb was to appear in the chart as a straight line of that bearing. In other words, "i" was the latitude to which φ would have to be "moved" in order to produce a chart that preserves true directions. Technically, "i" is the meridional part of latitude φ.[38]

Hariot's method was based on the observation that the radius vectors of the equiangular spiral were in a fixed ratio to each other when taken at regular angular intervals. From this he deduced that the length of a radius vector at a given point indicates the angle turned by the spiral up to that point. Hariot then proceeded to calculate the radius vectors of points projected from the 45° rhumb on the sphere. For any point on the circle of latitude φ on the sphere, this radius vector indicated the angle turned by the spiral, and hence the meridional part.

Hariot's method for the calculation of meridional parts was extremely sophisticated and precise by the standards of his time. It was perhaps too precise, since Wright's approach was quite sufficient for cartographic purposes, and there is indeed no indication that Hariot's figures were ever used in drawing a map. It is clear, however, that while the cartographic problem may have provided the impetus for Hariot's work, it was by no means its focus. Rhumb lines and the equiangular spiral became for him objects of inquiry in their own right, and he studied them in far greater detail than was required by the technical problem at hand.

In his papers entitled "De Helicis," Hariot calculated the length of the spiral, the area it encloses, and the sum of its radius vectors.[39] All calculations were made possible by his fundamental insight that the spiral can be approximated as an infinite series of similar triangles. These results were not to be repeated until thirty years later, when the Italian polymath Evangelista Torricelli took on the study of the equiangular spiral. Since the radius vectors were all in a fixed ratio to each other, Hariot could calculate their sum by simply summing an infinite geometrical series.[40] The sides of the triangles, which approximated the curve of the spiral itself, were also at a fixed ratio to each other. Using the same technique for summing up infinite geometric series, Hariot arrived at an accurate figure for the length of the approximate spiral composed of similar triangles. He followed this with a correct evaluation of the area enclosed within the approximate spiral, based on the fact that the areas of similar triangles also form a geometric series.

Once in possession of these results on the approximate spiral, composed of similar triangles, Hariot moved on. He posited that the fixed radial angle θ (equal to 1' in his rhumb calculations) tends to zero, resulting in a smooth true spiral composed of infinitesimally small triangles. This enabled him to accurately determine the same results for the true spiral that he had already derived for the approximate one.[41] In all these cases, Hariot arrived at his results through his insight into the inner structure of the curve as composed of an infinite series of discrete lines, bearing a fixed ratio to each other.

"THE CLUB OF HERCULES":
HARIOT'S SPIRAL AND THE CONTINUUM

So far, Hariot's investigation of rhumbs has followed much the same course as his study of refraction. In both cases, Hariot was confronted with a confusing and paradoxical situation. The ray of light was both reflected and refracted by the material medium, and in general, its behavior was not consistent with accepted predictions. The rhumb line was paradoxical and unintelligible, as the leading navigators of the age acknowledged. The smooth surface of the optical medium and the curved, rounded outline of the rhumb withheld their secrets tenaciously and seemed impenetrable at first glance. Hariot, however, was not discouraged and proceeded to scrutinize the inner structure of both the surface and the curve. The optical medium, he found, was in fact composed of discrete atoms with passages in between; the rhumb was similarly made up of an infinite number of distinct segments. By looking into the hidden inner structure, Hariot was able to solve both mysteries to his own satisfaction. The voyage of discovery into the secrets of nature culminated, in both cases, with the division of seemingly continuous objects into their fundamental components.

Hariot was well aware that his solution to the paradox of the rhumb was extremely problematic. The heuristic assumption that the equiangular spiral (the rhumb's projection onto the equatorial plane) was composed of an infinite number of similar triangles was satisfying to Hariot's materialist view of mathematical objects, and it proved effective for calculation purposes. Unfortunately, it ran contrary to the ancient principle, based on the paradoxes of Zeno and the problem of incommensurability that a continuous magnitude could not be composed of an infinite number of distinct parts. This truth of ancient mathematics was so fundamental that even mathematicians such as Archimedes who utilized infinitesimals always reworked their results to accord with the established Euclidean forms. The ancient "method of exhaustion" was developed precisely in order to circumvent the need for infinitesimals in geometrical proofs.

Hariot doubly violated the ancient rules in his calculations. First, in determining the length (or area enclosed, or sum of radius vectors) of the approximate spiral, composed of the sides of triangles, he assumed that this total length was the sum of an infinite number of finite straight lines. The assumption that a finite magnitude could be composed of an infinite number of finite sections ran clearly against the paradoxes of Zeno. Second, in calculating the length (or area enclosed, or sum of radius vectors) of the true spiral, Hariot assumed it to be composed of an infinite number of infinitesimal triangles.[42] In addition to having the same problems with Zeno's paradoxes as the first stage, this also encountered the problem of incommensurability, since it divided the line into its "atomistic" components.

Hariot's manuscripts show his concern about the paradoxical implications of his solution to the problem of rhumbs. The answers he proposed for these problems, however, were very different from the ones adopted by his ancient predecessors. Eudoxus and Archimedes attempted to avoid the problem by developing a consistent logical method that could prove the required results, while deliberately circumventing the infinitesimal nature of the problem. That was the purpose of the method of exhaustion. For Hariot, such an approach would be self-defeating: the whole purpose for his preference for the analysis of the rhumb over the addition of secants was that his method enabled him to "look inside" the fundamental structure of the rhumb. A method that would deliberately avoid doing so was worse than useless for Hariot's purposes. What was needed was a true understanding of the continuum, not a systematic avoidance of it.

Hariot's attempts to resolve the paradoxes plaguing his equiangular spiral are scattered throughout his unpublished manuscripts. Due to the hopelessly confused order in which the manuscripts were bound in the nineteenth century, there is no way to determine which folios belong together in single systematic studies and which should be considered separately. It is clear, however, that in numerous places in his papers Hariot joined his technical considerations of the inner structure of rhumb lines with subtle meditations on the nature of the continuum in general.[43]

The main focus of Hariot's concerns at this point seems to have been Zeno's Achilles paradox.[44] As Hariot was quick to discern, the Achilles dealt essentially with the summing up of an infinite geometric series. If we suppose, for example, that Achilles is ten times faster than the tortoise, and the original distance between them is "1," then by the time he reaches the tortoise's starting position, the latter had already advanced one-tenth that distance.[45] By the time Achilles advances to that point, the tortoise had already moved another one-hundredth of the original gap; and when Achilles had covered that distance as well, the tortoise will already be one-thousandth of the original distance beyond that point. Achilles would only catch up with the tortoise (if at all) at a point d determined as

$$d = 1 + \frac{1}{10} + \frac{1}{100} + \frac{1}{1000} + \frac{1}{10000} \ldots$$

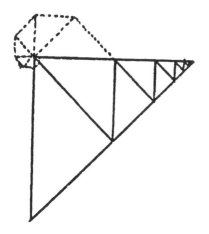 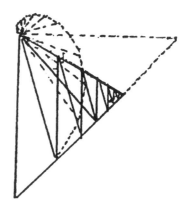

Figure 19.   Thomas Hariot, "In Achille." British Library Add. MS 6784, ff. 246–47. Similar graphics can also be found in British Library Add. MS 6784, ff. 248, 430, and British Library Add. MS 6785, f. 437 (Courtesy of the British Library).

of the original distance between the runners. In other words, d is determined as a sum of an infinite geometric series. This, of course, was precisely the fundamental problem with Hariot's analysis of the equiangular spiral: the sides, radius vectors, and areas of the infinite number of similar triangles were all in a fixed ratio to each other and therefore formed a geometric series. Hariot then proceeded to sum them up in determining the properties of the sphere. The Achilles, Hariot concluded, may have much to teach him about the true nature of rhumbs.

Hariot's attempts to reconcile his analysis of the spiral with the Achilles paradox can be glimpsed in various places in his papers. In one place, under the heading "In Achille," Hariot calculates the sum of

$$1/10 + 1/100 + 1/1000 + \ldots$$

and concludes that it equals

$$1111\ldots/100000\ldots = 1/9.$$

The meaning of this for Zeno's paradox is that Achilles would catch up with the tortoise after the latter had traversed one-ninth of the original distance between the two.

On the same page, Hariot appears to be translating the various relations between the distance covered by the two competitors and the time elapsed, into internal relationship within a triangle.[46] Hariot's true purpose here is revealed by the diagrams appearing directly above the calculations and in the adjoining folios. Each of the sketches shows a triangle divided into a series of smaller triangles, all similar and bearing a fixed ratio to each other (see Figure 19). All the smaller triangles together form the larger one, but when arranged differently, as some of the sketches show, they also comprise Hariot's approximate equiangular spiral. The number of the small triangles is potentially infinite, since there is no limit to the number of times the larger triangle can be divided in this systematic manner. Nevertheless, their combined area is clearly finite, since it is equal to the area of the larger triangle. This, of course, is also precisely the case with Hariot's approximated spiral, composed of these triangles, infinite in number, and yet finite in the area it encloses. The same, of course, is true of the length of the curve and the sum of its radius vectors, which can be calculated as the sum of the infinite geometrical series of the appropriate sides of the triangles. The meditation on Zeno's Achilles paradox here is designed to closely examine the paradoxical issues involved in Hariot's analysis of the equiangular spiral.

Another set of notes, from a different part of the Hariot collection in the British Library, reveals much the same pattern.[47] On one side of the page, Hariot jots down his speculations concerning the Achilles paradox, and sketches the respective courses of the two competitors. He supposes that A moves 100 miles in an hour, and B only 10 miles in an hour. He then constructs a table of their respective progress, assuming that B starts the race 100 miles ahead of A: when A had moved 10 miles, B is already at the 101 mile point and the distance between them is 91 miles; when A had moved 20 miles, B is at 102 miles, and the distance has shrunk to 82 miles.

Hariot continues this simple table until A had safely passed B after traversing 120 miles. The purpose of these apparently trivial calculations appears on the reverse side of the same folio: here Hariot again offers a sketch of his approximate spiral, composed of a decreasing geometric series of similar triangles. By analyzing the gradual convergence of the competitors in the Achilles, Hariot is attempting to clarify the behavior of the infinite but ever converging equiangular spiral.

Hariot summed up his difficulties by sketching the two hemispheres of the globe and a rhumb line stretching between them from one pole to the other.[48] He explains, "The probleme of a rectlinear rumbe infinitely turning about the center & yet the line finite & to be moved in an houre or any other given time to be made double to passe to two centers." How can a line composed of infinite parts, "infinitely turning about the center," be finite? How

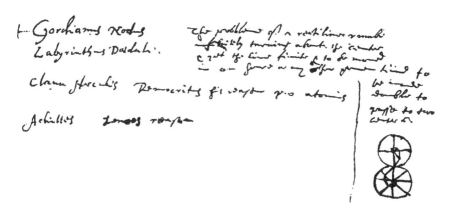

Figure 20.   Thomas Hariot's paradox of the rhumb. British Library Add. MS 6786, f. 349v (Courtesy of the British Library).

can such a curve be traversed in a finite and calculable time? And how is it possible that such a line, composed of an infinite number of discrete parts, can simply be doubled? These are precisely the type of questions that Zeno's paradoxes bring forth. The problem appeared to Hariot to be inscrutable and obscure, for he jotted down next to it: "*Gordianus Nodus. Labyrinthus Daedali*" (see Figure 20).[49]

In comparing the paradox of the rhumb to a labyrinth and Gordian knot, Hariot was taking part in a wider discourse of scientific and mathematical discovery.[50] The metaphor of the labyrinth had served his younger contemporary, Sir Francis Bacon, in describing man's relation to nature: "The universe to the eye of the human understanding," he wrote, "is framed like a labyrinth, presenting as it does on every side so many ambiguities of way."[51] For William Oughtred, years later, it stood for the obscurity of ancient mathematics, when unaided by proper symbols. His *Clavis Mathematicae* was to serve the reader "as it were Ariadne's thread, to guide them through the intricate Labyrinth of these studies."[52] Most importantly, for Edward Wright, Hariot's colleague in the quest for reforming marine cartography, the old plane chart was "much like an inextricable labyrinth of error, out of which it will be very hard for a man to unwind himself."[53] For Hariot, the rhumb line, representing a ship's course on a fixed bearing, was itself a labyrinth. It was confusing, impenetrable, and seemingly insoluble; a Gordian knot indeed.

It was, however, a central tenet of the narrative of exploration that Gordian knots will be cut, and the secrets of labyrinths will be exposed. Bacon's description of the world as a labyrinth came as a prelude to his account of the true method for exploring it. Oughtred's book was meant to be an "Ariadne's thread," which would guide the reader through the mysteries of the labyrinth; and Wright was, in fact, promoting his own reform of the plane chart, which would eliminate the "labyrinth of error." Hariot himself had also followed the same story line in his theory of refraction: the narrow portals of nature's mansion were, in the end, penetrated by the atom, and its secrets were revealed. The labyrinth of the rhumb was no exception. Hariot fully intended to break through its paradoxical surface and explore its true structure in detail.

Hariot hints at his method for penetrating the labyrinth of the rhumb by jotting down, just below his pronouncement of the Gordian knot: "Clava Herculis, Democritus his reason pro Atomis; Achilles, Zeno's reason."[54] Democritus' atomism, Hariot suggests, will serve as "The Club of Hercules," smashing through the mysteries of the labyrinthine rhumb.[55] The mention of Zeno's Achilles paradox lends credence to this view: by arguing that the assumption of infinite divisibility of magnitudes leads to absurd results, the Achilles seems to support the Democritean position that the continuum is, in fact, composed of atoms.[56]

Hariot does not argue his case here, or deal with the difficult problems involved in an atomistic mathematical approach. He does, however, utilize the basic elements of the tale of exploration in hinting at the solution to the paradox: the rhumb is presented as a confusing and imponderable "labyrinth," withholding its secret and defying mathematical analysis. Hariot then breaks through the difficulties by swinging the Club of Hercules. The smooth, impenetrable rhumb is transformed in the process into a collection of discrete, atomic segments. Much like the distant shores of America, or the problematic medium of refraction, the confusing, paradoxically smooth curve is mastered by carving it up into discrete fundamental parts.

"THE MYSTERY OF INFINITES"

Hariot had begun his investigation by considering a straightforward cartographic problem: By how much should a given latitude be displaced in order to produce a marine chart that would preserve true directions? His solution to the question involved the construction of an equiangular spiral, by approximating it as an infinite geometric series of similar triangles. This impressive technical solution, however, proved troubling to Hariot, because it raised further problems: How could a smooth continuous curve be composed of an infinite number of discrete sections? Such a construction, he well knew, contradicted the ancient lessons drawn from Zeno's paradoxes and the problem of incommensurability. Nevertheless, Hariot was undeterred. Rather than attempting to accommodate or circumvent these problems (as his ancient predecessors did with the method of exhaustion), he challenged the paradoxes directly. Despite the familiar contradictions, he insisted, the smooth line is indeed composed of discrete atomic segments. Democritean atomism, for Hariot, held the key for the true understanding of the continuous curve.

By this point, however, Hariot has clearly moved well beyond the confines of his original question. The issues raised by Zeno's paradoxes, on the one hand, and Democritean atomism, on the other, were not limited to the construction of rhumbs or the equiangular spiral. They were, rather, questions that applied to the structure of the mathematical continuum in general. Is the smooth continuum composed of indivisible parts? Are these segments finite or infinite in number? Or, is a continuum perhaps always simply a continuum, and can never be dissolved into fundamental component parts? Those were the kind of foundational questions raised by Hariot's investigations.

It is not surprising, therefore, to find that Hariot offered a more systematic study of these problems than the cryptic comments found in his cartographic investigations. Indeed, interspersed among his notes on the paradox of the rhumb are direct references to his general treatment of the problem of the continuum. The Club of Hercules and Achilles, it turns out, are not general musings on the subject of atomism and the continuum. They are, in fact, references to specific sections in Hariot's most comprehensive investigation of the problem of the continuum—the short treatise entitled "De Infinitis."[57] The close examination of the equiangular spiral has here led Hariot to consider the fundamental issues of the structure of the mathematical continuum.[58]

Hariot begins his study "De Infinitis" with a quote from Aristotle, who was the starting point for practically all discussions of the subject in medieval and early modern Europe.[59] Under the heading "De Continuo" he wrote:

> Aristotle in the beginning of his 6th booke of his Phisicks, & in the 26th text of the 5th booke, defined those things to be Continuo of which the ends are one. And in the 22th text of the sayd 5th booke that: Things are said to touch, if their boundaries are together. Things are together, which are in one primary place. Separate, which are in distinct places. Now for the better explication of the meaning of the definitions as also of there truth, let us understand first two materiall cubes A&B to be separate, that is to be in diverse places extremes and all. [Hariot's original text is a mix of English and Latin, and the Latin phrases have been anglicized here. The full text as originally written can be viewed in the footnotes.] [60]

In the *Physics* itself, these definitions led Aristotle to the conclusion that the continuum could not be composed of indivisibles. As he explains in the beginning of Book 6, indivisibles do not have parts (by definition), and therefore do not have "extremes." It follows that they cannot possibly form a continuum. Putting this consideration aside, Aristotle further argues that since indivisibles do not have parts, they can only touch each other as wholes. If that is the case—then they will occupy the very same place, and, as before, would never form a continuum.[61]

Hariot's own notes on the continuum follow a path very different from Aristotle's. Although he starts out with the ancient master's definitions, his speculations quickly take him in a different direction. This is initially evident when Hariot attempts to elucidate Aristotle's argument by positing "two materiall cubes" (see Figure 21). Aristotle, of course, never insisted on materiality and never suggested that cubes should represent the metaphysical concept of "continuity." But Hariot characteristically wanted to "see" the inner structure of the continuum as it "really" exists and therefore found Aristotle's logical niceties unsatisfactory. In 1609, when Kepler had offered him an explanation of refraction in terms of essences and qualities, Hariot expressed his strong distaste for such argumentation. "If such reasons satisfy you, I wonder," he wrote, cavalierly dismissing Kepler's theory.[62] The same dismissive attitude is evident here as well. Hariot wants to observe the structure of the continuum from up close, and Aristotelian logic is clearly not the way to achieve this perspective.

Hariot's dissatisfaction with Aristotelian reasoning is made clearer elsewhere in "De Infinitis," in a section entitled "Ratio Achilles." Although relying on the *Physics* as his primary source for Zeno's paradoxes, Hariot was clearly unpersuaded by Aristotle's treatment of the issues they raise:

> There is a reason of Zeno in Aristotle (in the 6th booke of his Phisickes, text 78) which for the force it seemeth to carry is called Achilles. And for that cause, no doubte, is the name also Achilles used in the example to expresse the reason. The which because it is agaynste Aristotle's doctrine & for that it contayneth matter pregnant of greater consequence concerning the doctrine of infinites, it being there but briefely and obscurely set downe with an answere uncertayne: I think good to set downe more particularly and largely: with Aristotle's answere as he hath it in the place abriged as also at full according to his owne doctrine in other places. To the end that comparing one with the other the truth may appear & perhaps found otherwise to be, then yet hath been by the Peripatetickes either noted or observed.
>
> The proposition of Zeno is. The swift runner (runne he never so swiftly) shall never overtake the slow runner (runne he never so slowly).[63]

Hariot is here explicit in his critique of Aristotle's treatment of the issues. Aristotle's discussion, he claims, is brief, obscure, and unconvincing. The source of Hariot's frustration becomes clear once we review Aristotle's orig-

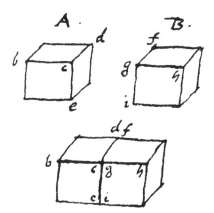

Figure 21. Thomas Hariot's depiction of the geometrical continuum composed out of square blocks, in "De Infinitis." British Library Add. MS 6782, f. 362 (Courtesy of the British Library).

inal comments on the issue in the *Physics*. After briefly explaining the nature of the Achilles paradox Aristotle wrote, "And the axiom that which holds a lead is never overtaken is false: it is not overtaken, it is true, while it holds a lead: but it is overtaken nevertheless if it is granted that it traverses the finite distance prescribed."[64] Elsewhere, Aristotle explains that the source of Zeno's error lies in his confusion of "infinity with respect to quantity" with "infinity with respect to division." Although it is not possible to cover the first of these in a finite time, it is possible to do so with respect to the latter. The Achilles, of course, deals with infinite division; therefore, there would be no problem for the fast runner to overtake the slower one.[65]

It should come as no surprise that both of these explanations appeared to Hariot fundamentally flawed. One may agree with the logical consistency of Aristotle's position or dispute it. But clearly, from Hariot's perspective, such explanations do not even begin to deal with the fundamental issues of continuity. How is it possible that a finite length be composed of an infinite number of finite segments, Hariot asked with regard to the nautical rhumb. Aristotle's discussion must have appeared to him quite unhelpful. For Hariot was not looking for a formal logical discussion: he was seeking the inside view of the continuum, which would reveal its hidden secrets.

Hariot left plentiful hints as to his views on the true structure of the continuum. In his discussion of the paradox of the rhumb, he referred to Democritean atomism as "The Club of Hercules" that would solve the difficulties. His repeated discussion of the Achilles paradox, which he clearly found to be persuasive, points in the same direction. The absurd conclusion that Achilles would never catch up with the tortoise followed from the initial presumption of infinite divisibility. The assumption must therefore be false, and the continuum must be formed of distinct atoms.[66] The sketches that Hariot appended to his discussion confirm this view. They show continuity as composed of distinct sections joined together, clearly displaying an atomistic conception of the structure of the continuum.[67]

In the "De Infinitis," however, Hariot seems in no hurry to state such determinate conclusions. Before arriving at any definite solution he devotes many folios to posing the problem in many different ways. He elaborates the paradoxes of the continuum over and over again, drawing on numerous eclectic sources: some are classical paradoxes, drawn from Aristotle, others are borrowed from the schoolmen of the fourteenth century, and still others were apparently devised by Hariot himself.

The strategy of presentation should be familiar by now, as it accords well with the narrative of exploration. The overseas explorers made sure to elaborate the imposing defenses of the land prior to announcing their own discovery of the way to break through. Hariot himself explained the confusing behavior of the optical medium at length before declaring his solution that matter is composed of atoms and passages between them. And Edward Wright wrote extensively about the "inextricable labyrinth of error" of the plane chart, before offering the addition of secants as his solution to the problem. All were, in their own way, following the guidelines of the standard narrative of discovery: the hidden mysteries of the interior, in this tale, must be well protected by strong and confusing defenses before the enterprising explorer penetrates all obstacles and finds his way to fame and fortune. The elaboration of the paradoxes of the continuum, therefore, fulfills an important role in Hariot's analysis of the continuum and its presentation. It is worthwhile to analyze at least a few of these discussions closely and in some detail.

One typical example can be found in folio 365, where Hariot elaborates a paradox of infinity, based on a discussion in the first book of Aristotle's *De Caelo*.[68] The paradox, which in its own way draws on the basic argument of Zeno's paradoxes, served Aristotle in arguing against an infinite universe. Simply put, the paradox discusses an intersection of a revolving line and a fixed infinite line and arrives at a contradiction. For Aristotle, this was simply an argument for ruling out such an interaction, thus concluding that the spherical universe must be finite. Hariot's approach was different: rather than simply ruling out the paradox, he analyzed the structure of the mathematical bodies in an attempt to reveal the source of the contradiction. Perhaps, he suggested, this will provide some clues as to the true structure of the continuum.

Under the heading "That in a finite time an infinite space may be moved" (see Figure 22), Hariot wrote:

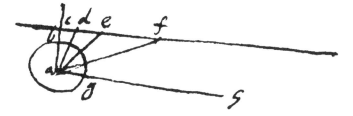

Figure 22.   Thomas Hariot, "That in a finite time an
infinite space may be moved." British Library Add. MS 6782,
f. 365 (Courtesy of the British Library).

Suppose the line *cef* at ultra to be infinite, & the line *ab* suppose to revolve &
describe a circle in a finite time, from *b* towards *g*. *ab* doth first respect *c*, then
d, after *e*, and so forth successively, no point in the infinite line is unrespected
by that time the line *ab* cometh to *ag* where the line is parallell & cutteth not
the former line infinite.[69]

The problem here is that the revolving line *ab* must intersect the fixed one
cf at each and every point of its infinite length before completing the simple
revolution from *ab* to *ag*, where it will be parallel to *cf*. This difficulty was
sufficient reason for Aristotle to conclude that the revolving line will never
be separate from the infinite fixed line, and will therefore not be able to complete the circular path.[70] Hariot, however, was only beginning his analysis:

The line *ab* having moved till he came to be *ah* that is parallell to *cf*. & so yet
continuing his motion of revolution: The lines are parallell but in one instant.
They never cut at an infinite distance, but at that instant are parallell.
   And if they cut them, they must cut at utrasque partes [that is, all parts] &
there being right lines there must be no space between them, but there distance,
by supposition is moved [by] the line *ab*. which implieth contradiction.

Hariot is here arguing that the revolving line *ab* does not cut the fixed line
cf when it has arrived at *ag*, since that would imply that two parallels are in
contact, from which it necessarily follows that they are in contact throughout their entire length. This, however, contradicts the assumption that they
are separated by the length *ab*, leading Hariot to conclude that they are not
in contact anywhere at that time. By contrast, he continues,

There must be a cutting at an infinite distance or else all the poyntes of the infinite line could not have been respected. & if that be not, some part of the infi-

nite line, that is some quantity finite only be cut. & that is at a finite distance; & then it macke an angle of quantity at the greatest distance of such cutting.[71] From that cutting the line by motion came to be parallell; That motion is made in an instant or in time. If in time, then in half the time the cutting must be further than the supposed furthest. If in an instant, one line wilbe in two places in one instant. Qua implicant.

The last point of intersection between the two lines, Hariot argues, cannot be at a finite distance. If it were, then the transition from intersecting to parallel lines would occur either in a finite length of time, or in an "instant" (by which he means "no time at all"). Either possibility, Hariot shows, would lead to a contradiction. He then continues, "The lines must therefore cut at an infinite distance before they come to be parallell. And that must be in time before or in an instant before. If in time, then in half the time they cut at greater distance than infinite or are parallell before they are parallell. Which both do imply contradictions." In other words, assuming the lines intersect at an infinite distance before becoming parallel, the movement from the last point of intersection to the parallel position can be accomplished either in a finite length of time or in an instant. If in a finite time, then in half that period the lines would either intersect at a further point or already be parallel. Both options contradict the assumptions, and Hariot moves on to consider the possibility that the movement happens in an "instant." He writes, "If in an instant before; then two instants are one or different. If one—implicat. If two there must be no other betwixt them. And then there should be a time greater then an instant & lesse then any time of quantity, that is indivisible, that is agayne, indivisible into partes of quantity. & so the like of poyntes etc."[72] The instant when the lines intersect and the instant when they are parallel can either be one and the same or two different ones. If they are the same, then as before, the line would be in two places at once, which is a contradiction.[73] If they are two separate instances, then there can be no other instant between them (otherwise it would amount to a finite span of time). That time between them must, therefore, be greater than an "instant" (which for Hariot means no time at all), or else the two instances would not be separated and be one and the same. It must also be smaller than any stretch of time, or else a real span of time would intervene between the two instances. This intervening time, according to Hariot, is simply a "time indivisible."

Hariot has taken the ancient paradox and stood it on its head. Aristotle was primarily concerned with logical consistency: since the assumption of infinity led to a contradiction, he simply excluded that hypothesis from the scope of his philosophy. Hariot, in contrast, far from excluding infinity, was attempting to understand it by reconstructing a close view of the continuum. The paradox served him not as an end point of his considerations, but as a tool for investigating the true inner structure of extension and time. In other words, if Aristotle was constructing a self-consistent logical system, Hariot was trying to unveil the secrets of continuity as they exist in the world.

Hariot was clearly fascinated with the paradox he elaborated: he seemed to take pleasure in recounting the various options available for solving the paradox, each in turn terminating in a contradictory dead end. As one reads through the notes, it becomes clear that for Hariot the continuum was indeed a labyrinth, where each promising path in succession is eventually found to lead nowhere. In the end, only one trail seems to promise a way out of the insoluble maze—the path of atomism. His final comment that "so the like of poyntes etc." shows that his argument about the time continuum applies to geometrical magnitudes in general. Only by dividing the seemingly smooth surface into its component indivisible parts can a way be found through the morass of the continuum.

Although Hariot's discussion of the *De Caelo* paradox ends in a suggestion of a possible solution to the Gordian knot, other sections in the "De Infinitis" seem to lead nowhere at all. Such, for example, is Hariot's discussion in folio 369 of the circle paradox, commonly attributed to the thirteenth-century scholastic Duns Scotus.[74] In the previous example Hariot starts by assuming infinite divisibility, only to end up with atomism; in this instance he presupposes the atomistic structure of the continuum and examines where this assumption leads him:

> Now will I propound some difficulties to be considered of. Seeing that every line is compounded ex atomis, & therefore [in] the perifery of a circle one atomis is succeeding one another infinitely in such manner as that the perifery is at last compounded and made. Now also seeing that the whole superficies is compounded ex atomis undiquoque sitis about the point *a* [that is, "located on all sides about the point a"] so many times infinitely, & to that number is then infinitely, till the circle supposed be accomplished. (See Figure 23.)

The surface, Hariot states, is composed of successive concentric circles around the center *a*, each composed of atoms and each touching its neighboring circles. "I demande then what wilbe the number of atomi that are deinceps about the poynt *a*. Infinite they must needes be, or else infinite lines could not be supposed actually from the poynt *a* to the perifery. And infinite are also in the perifery."[75] The circles making up the surface must each be composed of an infinite number of atoms, or else lines could not everywhere be drawn from the center to the circumference of the circle.

Figure 23. Thomas Hariot's paradox of the circle. British Library Add. MS 6782, f. 369 (Courtesy of the British Library).

But now I demande whether they are aequally infinite or not. If about the center are lesse infinite then there cannot from the center *a* to every poynt in the perifery be understood a right line. But we must understand those atomi about the center, that we supposed indivisible, divisible, which were absurd. . . . Neither can there be more because they being deinceps, one more cannot be betweene, there being no distance. & if there might be one lesse; there lacketh of the supposed actual & definite, & positive number although infinite.

Hariot here offers two arguments demonstrating that the infinite number of atoms on the inner circle is equal to the infinite number on the circumference. His first contention is that this equality is necessary for drawing straight lines from the center to the circumference. If there are fewer atoms in the inner circle, they would be divided by some of the radiating lines; which is, of course, absurd. His second argument has a distinctly materialist flavor: the number of indivisibles must be equal because each circle touches the ones next to it. If the larger circles were composed of a greater number of atoms, their extra atoms would have to be fitted in spaces between the atoms of the smaller circles. This, again, runs counter to the supposition that the circles are composed of indivisibles in contact. He then concludes, "Then I say in a great place where there could be no more or lesse in a lesse place, there are an aequall number."[76]

As he promised in the beginning of the paradox, Hariot had indeed "propounded some difficulties." His conclusion that any surface, regardless of its size, contains an equal (although infinite) number of atoms is perplexing and counterintuitive. This is especially the case given that Hariot is attempting to construct a materialistic model of the continuum, as composed of an "actual & definite, & positive number" of atoms. Three centuries later the mathematician Georg Cantor, who laid the foundations of set theory, would be satisfied with a similar line of reasoning. For Cantor, infinite sets are considered to be of the same order if a one-on-one correspondence can be established between their members. For Hariot, who was attempting to recon-

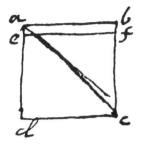

Figure 24. Thomas Hariot's paradox of the square. British Library Add. MS 6782, f. 369 (Courtesy of the British Library).

struct the "real" composition of the continuum, such a conclusion was problematic indeed. For try as one might, it seems impossible to visualize how surfaces of different magnitudes could be composed of the same number of indivisible parts.

The confusion of the labyrinth of the continuum grows thicker with every new example, and there appears to be no way out of the bewildering maze. Hariot, however, is not yet ready to extract himself from the morass. On the very same page in which he discusses the paradox of the circle he adds one more puzzle, further compounding the confusion surrounding the continuum. This "paradox of the square," like the previous one, was familiar to medieval scholastics and was elaborated by Duns Scotus (see Figure 24).[77]

> An other difficulty riseth from the square. If a line be compounded ex atomis, the diametrall line wilbe found to be aequall to the side. For suppose the line *ab* to be drawne from the poynt *a* of the line *ad* to the poynt *b* of the line *bc*. Then from the next poynt *e*, which is deinceps to *a* in the line *ad*, drawe a line to *f*, the next poynt to *b* in the line *bc*. So likewise from every next poynt in the line *ad*, to every next poynt in the line *bc*. Now the lines so drawne must needst be the least and most that may be, because they are deinceps & all. & they all cut the line *ac*. & of the line *ac* there can be no poynt betwixt two of the former lines, because they are deinceps. And therefore the number of poynts of the line *ac* are aequally infinite to the poyntes of *ab* etc. Per consequence the lines *ab* & *ac* [are] aequall.[78]

The assumption that lines are composed of indivisibles here leads to the absurd conclusion that the diagonal of a square ("diametrall," in Hariot's terminology) is equal to its side. For if we trace lines from each indivisible on one side to its counterpart on the opposite side, the argument goes, then they would all cut the diagonal at a regular indivisible distance from each other. If the diagonal contained more atoms than the sides, it would have been possible to draw lines through them, thus further dissecting the indivisible distance between the original parallel lines, which is absurd. The diagonal, then, contains the same number of indivisibles as a side, and they must, therefore, be of equal length.

There is no need to cite further instances of Hariot's fascination with the paradoxes of the continuum. Many other examples can be quoted both from the primary tract "De Infinitis" and from other parts of his manuscripts.[79] Repeatedly Hariot elaborates the inherent difficulties and contradictions latent in the deceptively simple concept of the continuum. The simple vision of a smooth surface becomes ever more obscure, confusing, and ultimately indecipherable. Hariot seems satisfied to pile the paradoxes one upon the other, until it truly seems that the nature of the continuum is impenetrable. Whether he posits an infinitely divisible continuum or one composed of indivisible atoms does not matter. Whichever route one takes in "De Infinitis," one will end up in a paradox, a contradiction, a dead end. Hariot's continuum is indeed a labyrinth, and he labels its inner paradox accordingly: "the mistery of infinites."[80]

But just when it seems that the "mystery" would never be solved, when it appears that there is no way out of the labyrinth of the continuum, Hariot offers a ray of hope. At the end of his paradox of the square, he notes, "But this difficulty wilbe made more playne by the next following, whereby wilbe found the meanes for the solution of all."[81] There is, Hariot states, a solution after all. There is a way through the confusing intricacies of the continuum, a path that will penetrate its secrets and solve the mystery.

It is, unfortunately, difficult to determine which text Hariot was referring to as "the next following." The confused order of the manuscripts leaves little hope that what was "next" for Hariot is still "next" in their present binding in the British Library. Indeed, the two folios directly following Hariot's optimistic pronouncement (folios 370–71) are filled with problems and questions of the familiar sort, rather than definitive answers. The next folio after that (folio 372), however, appears more promising, for it begins with the bold heading "Ratio Clava Herculis."

The Club of Hercules was Hariot's reference to Democritus' argument in favor of atomism.[82] It was to be the answer to the difficulties of the equiangular spiral, and it was the inevitable conclusion from Hariot's favorite paradox—the Achilles. The "Ratio Clava Herculis," in other words, was the long-promised solution to the perplexing difficulties of the continuum, Hariot's crushing rebuttal of the long list of problems and contradictions that he had elaborated so faithfully. It is the greatest irony of Hariot's work, and perhaps a measure of the difficulties he encountered, that the folio entitled "Ratio Clava Herculis" is the only page in the entire treatise that remained completely blank.

The probable argument of the "Ratio Clava Herculis" can be recovered from Roger Bacon's *Opus Tertium*. Bacon's main purpose in the relevant passages is to elucidate Aristotle's views on the continuum. Before proceeding with his argument, however, Bacon found it necessary to counter the atomistic reasoning of Leucippus and Democritus, which on account of its strength and subtlety he calls "The Club of Hercules." According to Bacon, Democritus reasoned that since a continuum can be divided anywhere, it can be divided next to any given point. What is true of one point is true of all, he argued, and the continuum can therefore be divided next to all points. But since a body is composed of that into which it is divided, it follows that it is composed of indivisible points or atoms.[83]

If this was indeed the thrust of Hariot's argument, it may not be so surprising that, at the last minute, he declined to write it down as his final and ultimate solution. Although Roger Bacon found it to be a formidable obstacle, which had to be demolished before Aristotle's authority could be established, to Hariot it may have appeared less than convincing. After promising the "meanes for the solution of all" paradoxes of the continuum, Democritus' argument appears decidedly anticlimactic. When we consider the severe problems which Hariot freely elaborated, we must conclude that Democritus' "reason pro atomis" falls short of its pretension to be an all-conquering Club of Hercules. Nevertheless, for Hariot, atomism was clearly the ultimate solution for the problem of the continuum: division into indivisible component parts was the only way to break through the labyrinth of the paradoxical continuum.

Hariot expounds on his solution in several places in the "De Infinitis." In his discussion of the *De Caelo* paradox, he concludes that time must necessarily be composed of "time indivisibles," and he extrapolates from this insight to geometrical points as well. In a different vein, more closely associated with his work on the equiangular spiral, he attempted to draw general conclusions from infinite geometrical series. In a famous passage in folio 363, which, according to Hilary Gatti, was inspired by Giordano Bruno, he writes:[84]

> In progressions that be infinite be they increasing or decreasing there are these passes. First to a quantity that beareth no rate to the first quantity geven, or rather because betwixt positive quantities there is a positive rate, I may call this rate infinite either in greatness or litleness according to the progression, in respect to the first quantity geven. Yet in respect of the progression following it is divisible or multiplicable till the progression being infinite hath for his second passe also a quantity of an infinite rate. Which is not only infinite in respecte of the first quantity of the last progression; but infinitely infinite in respect of the first in the first progression.[85]

So far, Hariot's meditations seem to lead him in the direction of the infinite divisibility of all magnitudes. The last item of a decreasing progression, he argues, may be infinitely smaller than the first item, but it is still infinitely divisible itself. For it is possible to construct another infinite series in which it would serve as the first (and largest) item. The same, he then argues, can be done to the last (and smallest) item of this second series. The reverse is true of increasing series, where the last and infinitely large item may serve as the first item of another increasing series. The process, Hariot explains, can be repeated for "a third, fourth & infinite other progressions & passes." Hariot here appears to be committed to the notion that infinity is relative to a given magnitude, rather than any absolute quantity. Aristotle and all other promoters of infinite divisibility would have been perfectly satisfied with Hariot's stance so far.

As in many other aspects of the continuum, however, appearances are deceptive. For following these noncontroversial and conciliatory statements, Hariot suddenly reverses his course and states boldly:

> And yet for a last in decreasing progressions we must needes understand a
> quantity absolutely indivisible; but multiplicable infinitely infinite till a quantity
> absolutely inmultiplicable be produced which I may call universally infinite.
> And in increasing progressions we must needes understand that for a last there
> must be a quantity inmultiplicable absolute, but divisible infinitely infinite till
> that quantity be issued that be issued that is absolutely indivisible.[86]

In the end, then, after endlessly repeated divisions, further division is no longer possible. "A quantity absolutely indivisible" is ultimately reached. The same is true of increasing series: after endlessly repeated multiplications, a quantity will be reached where multiplication is no longer possible.

Both the infinite and the infinitesimal are, for Hariot, specific magnitudes. He states, "That such quantity which I call universally infinite, hath not only act rationall, by supposition, or by consequence from mere supposition: but also act reall or existence; in an instant having a perfecte actuall being, or in time passed by motion both finite and infinite; with many reall consequences or properties consequent."[87] Indivisible magnitudes (as well as "universally infinite" ones) are actual entities for Hariot.[88] They exist in the world not as "mere suppositions," but "act reall or existence."[89] In spite of all the difficulties and contradictions, continuous magnitudes, according to Hariot, are actually composed of indivisible parts.

Hariot goes on to state his position in no ambiguous terms. Writing in Latin, in a few passages in folio 374v he makes the strongest possible claims as to the composition of the continuum:

Contacts are existence in two places with an indivisible distance.
The continuum is an aggregate of contacts.
The minimal continuum is an aggregate of two contacts.

And later:

The continuum is primarily composed of indivisibles in contact.[90]

Hariot has here deciphered the secret of the continuum to his own satisfaction. Despite the confusing intricacies and baffling contradictions surrounding it, the continuum is soluble after all. On close observation, its surface is revealed to be lined with innumerable breaks at regularly spaced intervals. The smooth, impenetrable continuum is finally mastered by breaking it into its essential component parts.

MAPPING THE MATHEMATICAL CONTINUUM

Hariot's investigation of the continuum in "De Infinitis" closely follows the basic structure of a geographical discovery tract. The basic questions that Hariot asks in his study seem far more suited to a geographical explorer than a traditional mathematician. A mathematician working in the Euclidean tradition would look to construct a coherent and consistent logical system, in which each necessary truth is derived from a previous, less complex one. This is clearly not Hariot's intention, for he seems perfectly content to elaborate paradox upon paradox and still, in the end, offer a solution that threatens to contradict them all.

Hariot's goal is to get a close view of what the continuum "really" looks like. He approaches it not like a geometer, relying on presuppositions and rigorous deduction, but like an explorer in an unknown land, carefully feeling his way through difficult terrain. His goal is to "see" the land as it is, "reall and existence." He has no interest in constructing a perfectly logical, but hypothetical world in the standard geometrical fashion. It is no coincidence that Hariot's end product is not a general system, in the manner of Euclid's *Elements*, but rather a materialistic sketch depicting the continuum at close range. In the end, like a true explorer, Hariot had constructed a map of the continuum, a drawing of "what it truly looks like," to guide future travelers through its confusing landscape (see Figure 25).

The "De Infinitis" tells the familiar story of the standard narrative of exploration. It starts with an enigma, "the mistery of infinites," which holds the key to great treasures of knowledge, such as Mercator projection charts and the properties of the equiangular spiral.[91] The secret of the continuum,

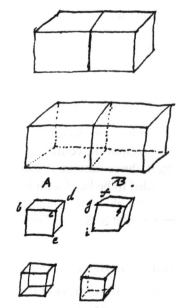

Figure 25.   Thomas Hariot's map of the continuum. British Library Add. MS 6782, f. 370v (Courtesy of the British Library).

however, is hidden in a confusing "labyrinth" and appears to the observer like a "Gordian knot." An abyss of darkness and confusion, Hariot demonstrates, surrounds the continuum on all sides: wherever one turns in search of its secret, one will get inextricably lost in a great maze of paradoxes.[92] The true nature of the continuum, like the refractive medium in Hariot's correspondence with Kepler, or unexplored lands of Virginia and Guiana, lies hidden behind confusing and seemingly impenetrable defenses. Hariot, however, does not despair. With the aid of Achilles and the club of mighty Hercules, he breaks through the confusing mists that surround the continuum.[93] The seemingly smooth surface of the continuum is found to be a series of discrete sections with indivisible spaces between them. These regular breaks offer access to the hidden secrets of the continuum and enable Hariot to possess its troves of knowledge.

Hariot's treatment of the mathematical continuum is, in the end, remarkably similar to his treatment of the coasts of undiscovered land or of the optical medium. In each instance hidden treasures were posited in a hidden, inaccessible "interior," where they were protected by a continuous barrier. All attempts to penetrate the obstacles and reach the hidden treasures ended in predictable failure. In each instance—geographical, optical, or mathemati-

Figure 26. Thomas Hariot, an unidentifiable coastline. British Library Add. MS. 6787 f. 473v (Courtesy of the British Library).

cal—Hariot's solution was the same. In Virginia, the formidable coast was found, on close observation, to be traversed by clear passages. These enabled the English fleet to break through the chain of islands and penetrate the golden interior. The seemingly smooth optical medium also effectively defeated all attempts at the explanation of refraction and shielded the secrets of nature's mansion from prying eyes. Hariot, upon examination, announced that "that which to the senses seems to be continuous in all its parts is in fact not so." The medium was, in fact, composed of atoms separated by vacuous passages through which light rays may pass.

In the case of the mathematical continuum, Hariot poses similar questions, tells a similar story, and offers the same solution: the mathematical continuum appears smooth, confusing, and impenetrable, defeating all attempts to decipher its hidden secrets. On close examination, however, it is found to be composed of discrete indivisibles with breaks in between them. This structure, in turn, allows Hariot to master the continuum and extract its secrets, as he does in the case of the nautical rhumb and the equiangular spiral. The geography of discovery, which shaped the explorers' maps of undiscovered lands, and Hariot's vision of the inner composition of matter, has here structured the vision of the mathematical continuum itself. Hariot, in essence, had drawn the map of the undiscovered land of mathematical continuum.[94]

A final example vividly demonstrates the congruence of Hariot's geography of undiscovered lands and his model of the mathematical continuum. Hariot's papers in the British Library contain several sketches of coastal maps.[95] One of them can be confidently identified as the island of Trinidad and northeastern Guiana.[96] Others cannot be identified: they all closely resemble each other in their generic construction of a coastline broken by regularly spaced openings (see Figure 26).

Figure 26 shows what a coastline looked like to Hariot; Figure 25 shows his vision of the mathematical continuum, composed of an aggregate of discrete segments. The resemblance is not coincidental. The geographical space of the foreign coastline and the geometrical space of the continuum were both structured by the Elizabethan narrative of exploration and discovery.

CHAPTER 6

Navigating Mathematical Oceans

EXPLORATION VERSUS MATHEMATICS

Giordano Bruno, the flamboyant Italian philosopher, did not think highly of mathematicians: "[They] are in fact like the interpreters who translate words from one language to another, but then there are others who fathom the meanings, and not the words themselves. The former are in fact like peasants who report the trends and patterns of a skirmish to a distant captain; they themselves do not understand the steps, the reasons, and the art by which they became victorious."[1] Even Bruno's hero, Copernicus, came under criticism for his excessive adherence to mathematics. Copernicus, "being more intent on the study of mathematics than of nature," Bruno charged, "was not able to go deep enough and penetrate beyond the point of removing from the way the stumps of inconvenient and vain principles."[2]

The study of mathematics, it is clear, was of limited use for Bruno. Mathematics can correctly describe certain aspects of the world, its objects, and the relations between them. It cannot, however, move beyond this. It will never know the true causes of things, their inner structure and composition, and what makes them appear as they do. It is forever confined to mere surface appearances, without hope of penetrating to the hidden secrets of nature.

Bruno followed his dismissal of mathematical learning with exuberant praise for a different approach to the pursuit of knowledge: geographical exploration. "The ancient Typhon," he wrote several passages later, "was to be praised for having found the first boat and for having crossed the sea with the Argonauts."[3] Similarly, "Columbus is to be celebrated as the one about whom it had been foretold long ago":

There shall come an age
In the far off years, when the Ocean
Shall unloose the bonds of things

And a big land shall emerge, while Tethys
Will disclose new realms and Thule
Shall no longer be the limit of dry land.[4]

Although Bruno had high praise for the great geographical voyages, they were, in fact, a mere prelude to the intellectual exploration undertaken by "the Nolan," as Bruno called himself in the dialogue after the name of his hometown. The Nolan, he claimed, has undertaken a grand tour of the universe:

> Now here is he who has pierced the air, penetrated the sky, toured the realm of the stars, traversed the boundaries of the world, dissipated the fictitious walls of the first, eighth, ninth, tenth spheres, and whatever else might have been attached to these by the devices of vain mathematicians. . . . Thus aided by the fullness of sense and reason he opened with the key of most industrious inquiry those enclosures of truth that can be opened to us all, by presenting naked the shrouded and veiled nature.[5]

The Nolan, in other words, is an intellectual voyager to the inaccessible regions of the cosmos, who presents its hidden secrets to an eagerly awaiting world. The true genius, for Bruno, is nothing but an explorer of the hidden realms of nature.

Bruno was not alone in hailing the voyages of discovery as the proper model for science. Earlier we noted how Bruno's younger contemporary, Francis Bacon, declared his intention to explore the "Intellectual Globe" in the same manner in which the terrestrial globe had been "laid widely open and revealed." Indeed, citing the great geographical discoveries as models for the exploration of nature was to become commonplace among promoters of the new science in the seventeenth century.

Significantly, however, Bacon shared with Bruno not only his admiration for seafaring but also his low opinion of mathematics and its potential role in the sciences. Famously, Bacon's empirical program for the reform of knowledge did not assign any clear role to mathematics and has been criticized for this "failure" by modern philosophers. Several decades later, when the analogy between voyages of discovery and the study of nature was widespread among experimentalists in the Royal Society, Robert Boyle was still wary of the application of mathematics to nature.[6] Many early modern natural philosophers, it would seem, were quick to adopt geographical exploration as a model for scientific discovery, while keeping a guarded distance from the rigors of mathematics.[7]

How are we to understand this suspicion of mathematics among the early modern reformers of knowledge? Its source lay undoubtedly in the fact that classical mathematics, as commonly perceived, shared none of the intellectual goals espoused by the reformers. Classical mathematicians in early modern Europe did not seek out new knowledge or uncover hidden truths in the manner of geographical explorers. Taking Euclidean geometry as their model, they sought to draw true and necessary conclusions from a set of simple assumptions. The glory of mathematics, for them, lay in the certainty of its demonstrations and the incontrovertible truth of its claims, not in the discovery of new and veiled secrets. Indeed, what could possibly be left "hidden" and "undiscovered" in a system where all truths were already implicit in the initial assumptions?[8]

This view of mathematics was expressed most clearly by Christopher Clavius, founder of the Jesuit mathematical tradition, in his tract "In Disciplinas Mathematicas Prolegomena," dating from the 1580s. Using the common medieval classification, Clavius divided mathematics into "pure" and "mixed" domains. Pure mathematics included arithmetic and geometry, and mixed mathematics roughly corresponded to our notion of "mathematical sciences." It was comprised of fields such as astronomy and music, as well as engineering and geography. These, according to Aristotle, were disciplines that drew on the results of pure mathematics and applied them to particular physical situations.

But whether mixed or pure, Clavius insists, the mathematical sciences "proceed from particular foreknown principles to the conclusions to be demonstrated."[9] "The theorems of Euclid and the rest of the mathematicians," he continues, "still today as for many years past, retain in the schools their true purity, their real certitude, and their strong and firm demonstrations." He then concludes, "So much do the mathematical disciplines desire, esteem, and foster truth, that they reject not only whatever is false, but even anything mere probable, and they admit nothing that does not lend support and corroboration to the most certain demonstrations."[10] For Clavius, as for many of his contemporaries, all forms of mathematics proceed by deducing undisputed truths from generally known and accepted first principles. They have little to say of the exploration and discovery of hidden and unknown realms of knowledge.[11]

It is hardly surprising that such an approach proved unattractive to the new reformers of knowledge who sought to fashion themselves as intellectual explorers. A system of knowledge in which all truths were logically derived from accepted postulates offered no hope of breaking the established intellectual mould. Classical mathematics, it was readily admitted, correctly described certain aspects of the world, its objects, and the relations between them. It could never, however, move beyond this. It could never investigate the true causes of things, their inner structure and composition, and what makes them appear as they do. It is forever confined to mere surface appearances, without hope of penetrating to the hidden secrets of nature.[12]

Francis Bacon complained that mathematicians "delight[ed] in the open plains of generalities, rather than in the woods and inclosures of particulars." This, he claimed, was "to the extreme prejudice of knowledge."[13] In a somewhat similar vein Giordano Bruno, who viewed himself as Columbus's intellectual heir, charged that mathematicians are "like the interpreters who translate words from one language to another, but then there are others who fathom the meanings, and not the words themselves."[14] As the case of Robert Boyle makes clear, the shadow cast by these views on the role of mathematics in the reform of knowledge was still in evidence in the latter part of the seventeenth century.

The critique of mathematics by leading intellectual reformers posed a serious challenge to mathematical practitioners.[15] After all, the analogy between the voyages of discovery and scientific investigations was one of the hallmarks of Renaissance science.[16] If, as Bacon and his colleagues claimed, the exploration of the unknown meant nothing to mathematics, what role could it play in the newly emerging order of knowledge? If it is indeed uninterested in "exploring the intellectual globe," then it may well find itself relegated to irrelevance in the new intellectual landscape.

The manner in which some mathematicians responded to this challenge is the main focus of this chapter. Many undoubtedly stayed their course, insisting on the logical consistency and necessary deductive reasoning that characterized their field. Others, however, joined in the critique of traditional mathematics. Repeatedly they voiced the complaint that all the ancients left them was a web of beautiful results, but no chart of the roads that led them there and no hint as to how to proceed with further discovery. In response, they opted for a new course: taking their cue from the experimental philosophers, they chose to adopt the rhetoric of exploration and make it their own. De-emphasizing the traditional role of mathematicians as the bearers of eternal unchanging truths, they instead fashioned themselves as explorers of the unknown, as intrepid travelers in the hidden lands of mathematics. Most importantly, this shift in the understanding of the role of mathematicians and their purpose brought about a change in the technical practice of mathematics.[17]

We have already observed how the close involvement of mathematicians in the culture of exploration in England contributed to the development of new mathematical approaches. Thomas Hariot, who was a geographical explorer as well as a pathbreaking mathematician, embodied the encounter between the two in a manner no other mathematician could match. His participation in the maritime ventures of his time supplied him with the subject matter for much of his mathematical studies, as well as a radical new vision of mathematics as a voyage of discovery. This profoundly informed his views on the nature of the continuum and his radical infinitesimal methods. But while no mathematician could rival Hariot's credentials as an explorer, others, both in England and on the Continent, were moved by a similar impulse. Inspired by the image of mathematics as a journey in unknown lands, they reconfigured their studies to conform to this vision.

In this chapter I extend my argument and trace the fortunes of the mathematics of exploration beyond the English context. The imagery of mathematics as a voyage of discovery was closely associated with the development of the new and controversial infinitesimal techniques. These methods were to dominate seventeenth-century mathematics and ultimately lead to the development of the calculus. Repeatedly, mathematicians who presented themselves as voyagers in search of hidden secrets also promoted the new methods of indivisibles. This correlation is not coincidental. Much as the image of the natural philosopher as an intellectual explorer went hand in hand with the new empiricist practices, the image of the mathematician as a voyager in strange seas supported a new kind of mathematical approach. It suggested a new set of possible questions as well as acceptable answers to practicing mathematicians. In doing so, the imagery of exploration helped promote the new infinitesimal methods.[18]

Needless to say, the narrative of discovery was not the only theme used by promoters of the new mathematics to describe their practices. Mathematicians also described their craft in terms of a "hunt" (venatio), an interrogation chamber, and a "royal road," and references were made to classical themes such "labyrinths" and "Ariadne's thread" as well as the unavoidable Argonauts. Each of these themes is derived from a distinct tradition, whether medieval, courtly, or classical; each carries with it a unique set of connotations derived from its source and its setting. Clearly, the tale of exploration and discovery was far from being the sole narrative used by infinitesimalist mathematicians. Rather, they drew on a wide array of cultural resources to define, legitimize, and promote their approach. The narrative of exploration should therefore be understood as part of a web of related metaphors, which sought to highlight and guide the search for new and previously unimagined knowledge. This chapter focuses specifically on exploration mathematics and its fortunes in seventeenth-century Europe, but it can also be read as a study of different strands of early modern mathematical rhetoric.

I fully acknowledge the wide variety of themes and metaphors used by mathematicians, but here I would like to make a case for the special status of the narrative of geographical exploration. While it was undoubtedly part of a wider set of narratives, it was nevertheless a crucial and indispensable part of this web. Unlike other themes, which drew their significance from their local context, the narrative of exploration was recognized throughout Europe as an emblem for the reform of knowledge and the search for new discoveries.

This was the case not only for mathematics: the advocates of the new natural philosophy, for example, employed a varied and colorful imagery to define and promote the reform of knowledge through the experimental study of nature. And yet, the vision of geographical discovery with its promise of undreamt new continents of knowledge held a unique and prominent place for them in their understanding of their project. It was a theme that was as useful and recognizable to Bacon in the Stuart court as it was to the wandering Giordano Bruno or Galileo in Medici Tuscany. Despite the remarkable differences in their standing, outlook, and locale, all of them used the rhetoric of exploration to describe their experimental search for the secrets of nature. Much the same can be said of the new mathematicians: although they employed a wide variety of tropes, depending on their aims and circumstances, the narrative of exploration was one that they found indispensable. To this they returned again and again, despite differences in their local settings and traditions.

How important was the narrative of exploration in the web of interrelated metaphors employed by the new mathematicians in early modern Europe? I leave it to the readers to draw their own conclusions. My purpose in this chapter is to draw attention to the continued prominence of this trope in mathematical circles beyond the confines of Elizabethan England and into the seventeenth century. Furthermore, just as in England it helped support the new mathematical approaches of Thomas Hariot and his circle, in the wider European context, it was closely associated with the infinitesimal techniques developed by Stevin, Cavalieri, and their colleagues.

A MARINER IN STRANGE SEAS:
THE NEW MATHEMATICAL IMAGERY

In 1583 the Dutch mathematician and engineer Simon Stevin introduced his *Problematim Geometricarum* with a poem by Luca Belleri extolling the virtues of mathematics:

> Truly, then, the ancients called
> divine mathesis that which by
> its craft enabled to recognize

the supreme seat, the ways of the earth and sea
and to see in person the hidden places in the dark
the secrets of nature.[19]

This is indeed a startling account of mathematics' nature and purpose, when compared with the traditional views expressed by Clavius. The mathematician is seen here as an explorer, navigating "the ways of the earth and sea" and viewing "in person" the hidden secrets of nature. This, of course, is precisely what reformers like Bacon and Bruno were seeking to achieve in their new models of knowledge. Dismissing the authority of old canons, they sought to gain knowledge of the world like explorers do—through direct personal experience. It was clearly not what mathematicians themselves had sought to achieve over the centuries. Indeed, as Clavius had argued, the strength of mathematics lay precisely in the fact that it was not dependent on personal experience or sense perception, but based strictly on pure and rigorous reasoning from first principles. In applying the rhetoric of exploration to mathematics, the poem proposes a fundamental shift in the understanding of the very nature of the field.

Stevin returned to the exploration metaphor in other works as well. In his dedication to *Dime*, his best-selling treatise on decimal notation, he writes, "But as the Mariner, having by hap found a certain unknown island, spares not to declare to his prince the riches and profits thereof, as the fair fruits, precious minerals, pleasant champions, etc. and that without imputation of self glorification, even so we may speak freely of the value of this invention."[20]

Stevin is here again an explorer in the uncharted lands of mathematics. Rather than promote his treatise as the result of rigorous mathematical deduction, he chooses to describe it as an "invention," the happy result of his mathematical travels.[21] Just like an unknown land, it is discovered through chance wanderings, and like it, it holds the promise of great riches.

Stevin, it should be noted, was not an academic mathematician: he was a high-level official in the court of Maurice of Nassau, responsible for digging canals, building dams, and constructing fortifications. We should not perhaps be surprised to find that he did not share Clavius's lofty insistence on the strictly deductive nature of mathematics. His views on the subject were likely to be far more utilitarian, concerned with achieving durable results rather than confirming the eternal truth of mathematical propositions. His description of mathematics as a voyage of exploration is perhaps less remarkable if we consider that as an engineer and administrator he was often concerned with finding innovative ways for the solution of practical problems.[22] The imagery of mathematical exploration, however, did not long remain the exclusive domain of practical men like Stevin. Early in the seventeenth century we find the same language being used in more academic settings on the opposite end of the Continent.[23]

In the 1630s and 1640s leading members of Galileo's circle began referring to their mathematical studies in terms of travel and exploration. Unlike in the Netherlands, Italian seafaring was on the decline in the seventeenth century, and Italian mathematicians were usually far removed from actual maritime ventures. Despite this, the rhetoric and imagery of the voyages flourished among Italian natural philosophers and mathematicians. Maritime voyages and physical experiments onboard ship figured prominently in Galileo's *Dialogue Concerning the Two Chief World Systems*.[24] He often referred to scientific work as unveiling the hidden secrets of nature and applied this vision to the study of mathematics as well.[25]

At the end of the first day of the *Dialogue*, for instance, Galileo explained that mathematical truths are "clouded with deep and thick mists, which become partly dispersed and clarified when we master some conclusions."[26] We are easily reminded here of arctic explorers such as Frobisher and Davis, seeking to penetrate the clouds and thick mists that obscure their way. Much the same tone is present in his communications with his disciple, Evangelista Torricelli, who was renowned for his mathematical work on indivisibles as well as for his famous experiments on atmospheric pressure. In congratulating Torricelli for his mathematical achievements, Galileo wrote that by using his "marvellous concept," he "demonstrates with such easiness and grace what Archimedes showed through inhospitable and tormented roads ... a road which always seemed to me obstruse and hidden."[27] The language is indeed suggestive. As before, we are in a land of marvels and secrets, clouded by thick mists, with only difficult and convoluted passages leading through to them. Torricelli is praised for breaking through to his "marvellous concept" and blazing a trail for others to follow. He is indeed a mathematical explorer.[28]

Torricelli himself uses the travel imagery more explicitly in a lecture on the nature of geometrical reasoning given in the 1640s. Comparing the relative virtues of the humanities and geometrical studies, he argues that although in the humanities one may encounter some truth on occasion, it is usually of little significance "and so much entangled in the mist of falsities that accompany it, that the speculative mind will have great difficulty in separating the shadows of fog from the images of truth." In contrast, "in geometry books you will see in every page, nay, in every line, the truth is laid bare, there to discover among geometrical figures the richness of nature and the theatres of marvels."[29]

Much like Clavius in the "Prolegomena," Torricelli here is clearly intent on preserving mathematics' traditional claim to clarity and certainty in contrast to the confused and contested nature of other fields of knowledge. The source of these unique features of mathematics, however, is radically different for Torricelli than it was for his predecessor. Geometry's superiority is not derived from its rigorous logical structure, or Clavius's "most certain demonstrations." Instead it resides in its ability to reveal the riches and marvels hidden among geometrical figures. For Torricelli, the geometer is one who explores and seeks out those hidden secrets and brings them to light. This is indeed a very different image of mathematics than Clavius's systematic elaboration of deductive truths from self-evident first principles.

Significantly, Torricelli frequently utilized exploration metaphors when referring to the work of Bonaventura Cavalieri, his friend and fellow mathematician. Cavalieri was known to his contemporaries and to future generations as the founder of the "method of indivisibles" and was therefore a close collaborator of Torricelli.[30] In his *Opere Geometriche*, for example, Torricelli reassures his readers that he does not intend to venture upon "the immense ocean of Cavalieri's indivisibles, but, being less adventurous, he will remain near the shore."[31] Elsewhere, he calls Cavalieri a "discoverer of marvellous inventions" and credits him with being the first to venture upon "the true royal road through the mathematical thicket . . . who opened and levelled it for the profit of all."[32]

Torricelli is here drawing on a famous classical tale: Euclid, according to tradition, admonished King Ptolemy I that "there is no royal road to mathematics," implying that there is no way around the arduous method of rigorous geometrical deduction.[33] Significantly, however, Torricelli uses the story to point out Euclid's error: there is indeed a "royal road" to mathematics, which circumvents tedious geometrical demonstrations, and Cavalieri has found it. He has paved a road through the difficult mathematical terrain, obfuscated by traditional geometrical practice, and opened the way to great marvels from which all will profit. The imagery here is familiar by now; it is almost identical to Galileo's description of Torricelli as one who opened a clear passage in place of the old tortuous and convoluted roads.

Cavalieri responded to his friend in kind and described Torricelli as the one who uncovers hidden gems. In a letter of July 10, 1641, he urges Torricelli to "divulge in print those treasures of yours, which should not remain hidden in any way."[34] Two years later he congratulates his colleague for the imminent publication of his works, stating that "it will be of great benefit to the scholars, which will enrich themselves with precious gems."[35] In other places in their extensive correspondence Cavalieri repeatedly refers to Torricelli's work as filled with amazing "marvels," "wonders," "precious stones," and "splendors."[36] This theme of searching for hidden gems and treasures has a long history, dating back to medieval times, and is not necessarily a part of an exploration narrative. As presented here, however, the trope is indeed reminiscent of our tale of discovery. In their own eyes, both Torricelli and Cavalieri were seekers of hidden secrets and gems, opening new passages through the difficult and confusing mathematical terrain.[37]

Galileo and his disciples, it appears, shared a fundamental vision of the nature of mathematics and its goals: great secrets and marvels, they held, lie within the mathematical fold, obscured by the fog and thicket of ignorance and confusion. The mathematician, like an explorer, must find his way through fog and wilderness and retrieve the elusive gems. Mathematics, for them, is a science of discovery: it is not about the systematic elaboration of necessary truths, but rather about the uncovering of secret and hidden "gems" of knowledge. Its goals have little in common with traditional Euclidean geometry; they have much in common with the aims and purposes of the newly emerging experimental sciences.[38]

For Galileo's circle the tale of exploration and discovery was a favored literary trope that helped shape their scientific practice. For their English contemporaries it was all that and more, for they were active participants in the British imperial project of their day. Thomas Hariot was a member of Raleigh's first Virginia colony of 1585, where he explored the Atlantic seaboard and reported his findings to his patron. After his return he provided continuous technical support for Raleigh's various imperialist ventures— drawing maps of distant regions, producing navigational instruments, and lecturing Raleigh's officers on their use. The effects of his involvement in the voyages on his mathematics was profound.[39] Like Stevin, he came to view his practice as a voyage of discovery in its own right, an exploration of the hidden secrets of mathematics.

Edward Wright was also actively involved in the voyages, as the title to his most famous work *Certaine Errors in Navigation* amply attests. He took part in the Earl of Cumberland's raid on the Azores in 1589 and was hailed as an adventurer in search of wealth and glory on the oceans of mathematics. Much like Stevin, Hariot and Wright viewed their mission as exposing the hidden secrets and mysteries of their field, rather than establishing incontestable truths. Both figuratively and literally, then, Elizabethan mathematicians were navigators and explorers, seeking gems and treasures on the high seas.

The death of Queen Elizabeth brought an end to the heroic age of English mathematics: seventeenth-century mathematicians were content to stay home rather than "put out to any foreign discoveries" (to use the phrase of one of the more prominent among them, William Oughtred).[40] Even in the absence of actual voyages, however, the vision of mathematics as a search for hidden secrets remained strong in England throughout the period. Such a view is hardly surprising in a country that adopted the Baconian exploration of the intellectual globe as part of its national ethos. Mathematics as exploration, after all, was closely allied with the experimental sciences: the aim of both was to explore uncharted terrain and bring to light hidden secrets and treasures.[41]

A prominent exponent of this view was John Wallis (1616–1703), Savilian professor of geometry and, along with Isaac Barrow, the leading mathematician in England in the generation before Newton. Wallis, it should be noted, belongs to a later generation than the other mathematicians discussed in this chapter. By the time he wrote his major works the use of infinitesimals was so prevalent among European mathematicians that it could not be seriously challenged. The nature, meaning, and proper use of infinitesimals were, of course, a matter of hot dispute, as Wallis's contentious exchange with the leading French mathematician Pierre de Fermat and his colleagues makes clear.[42] The actual use of infinitesimals, however, was no longer a novelty. The overwhelming majority of practicing mathematicians, even those who were hostile to the "exploration" view of mathematics, accepted infinitesimals by this time.[43]

Wallis is exceptional among our group in another way, because he never explicitly utilized the language of exploration in a mathematical context. He used related tropes, which brought out a similar vision of the meaning and purpose of mathematics. He did not, however, literally compare the mathematician to a voyager, in the manner of Stevin or Cavalieri. Despite these discrepancies, I have chosen to include Wallis in my survey because he clearly stands in the tradition of the mathematics of exploration and can provide clues as to its subsequent development.

Wallis openly declared his admiration for Cavalieri (which he knew through Torricelli's account of his work), Hariot, and Oughtred and viewed his work as the continuation of theirs.[44] Even more important perhaps was his close association with the English experimental community. During the Interregnum Wallis was almost certainly a member of Boyle's "Invisible College" and, in 1660, a founding member of the Royal Society. This may seem surprising for a mathematician, since the early Royal Society was far better known for its adherence to Baconian empiricism than for promoting mathematical excellence.[45] Wallis, however, was not an orthodox mathematician: he insisted, for example, that the highest form of knowledge is acquired by direct sensual perception, "tasting and seeing" in his words, rather than through intellectual abstractions.[46] This is, undoubtedly, an unusual position for a mathematician to adopt. Furthermore, his mathematical approach was shaped by what he referred to as "the method of induction," which consisted of the Baconian practice of drawing general conclusions from a limited number of tests. Wallis's mathematical approach, it may be said, was designed to present the hidden truths of mathematics to the senses by using the inductive method. It was a type of mathematics that closely approximated the experimental practices of Wallis's colleagues in the Royal Society.[47]

True to his experimentalist leanings, Wallis dedicated his 1659 treatise *De Cycloide* to his friend and colleague, Robert Boyle. In complimenting Boyle for his persistent pursuit of the secrets of nature, Wallis writes:

> In the hunt of the true philosophy . . . you pursue nature as if by iron and fire . . .
> you follow to the most hidden secret recesses, and penetrate as if to its visceral
> parts, that it is really a wonder if the prey does not give itself up to you. In any
> investigation you harass nature as if it is tied to a rack, harshly, or even cruelly
> I would say, by torture and more torture, that ultimately it leads to all the secrets
> being confessed.[48]

The imagery here is violent and vivid and draws on several narrative sources. The familiar hunt metaphor is intermingled here with suggestions of sexual conquest and the judicial interrogation of nature under torture. Despite the extreme rhetoric, however, Wallis so far remains very much within the Baconian tradition: the good experimenter seeks to hunt down the elusive secrets of nature by subjecting her to systematic unrelenting examination.[49] He then proceeds to extend this narrative well beyond its traditional bounds: "Similarly in mathematics, you turn analytical equations (the mathematical rack); for no other instrument, even the most subtle, can search everywhere for what is hidden, or extricate what is veiled."[50] The Baconian imagery of experimentation is here applied directly to mathematics. Like the experimental sciences, mathematics is concerned with shedding light on hidden secrets, and like them it proceeds by systematically exploring all that is "hidden" and "veiled." Even analytical equations, according to Wallis, are not simple expressions of universal logical relations, but instruments of torture, designed to extract the hidden secrets and treasures that mathematics conceals within its fold.

The imagery Wallis uses here is not, of course, that of voyaging and exploration. Wallis presents us with no forbidding oceans or undiscovered lands. Instead, he adopts two other of Bacon's favorite tropes—the image of the search for the secrets of nature as a hunt and the view of the natural philosopher as an interrogator of nature, determined to extract her secrets.[51]

Both sets of imageries were used extensively in this period to describe the new procedures advocated by the experimental philosophers, and both were often used in conjunction with the "geographical discovery" theme. When applied to mathematics, they implied a vision of the field that was very close indeed to the one promoted by "exploration" mathematicians. All three versions presented mathematics as withholding secrets, which must then be sought out and brought to light. All three, furthermore, suggested that this could not be achieved simply by deductive reasoning from first principles. What was required was an active pursuit, interrogation, or exploration of the unknown, which will reveal the inner secrets of mathematics.[52]

It was a radical new vision of mathematics, promoted not only in England but also in the Netherlands and in Italy. The gulf between this vision of mathematics and the classical view loomed wide. Traditionally mathematicians had emphasized the rigorous deductive nature of their field and the absolute certainty of its results. Mathematical results, in this view, may seem surprising to a novice encountering Euclidean geometry for the first time. For an experienced mathematician, however, the field held no mysteries: its basic results had been known since antiquity, and in any case they are already implicit in the initial assumptions. Mathematics, in this view, provided little that was new or surprising, but its conclusions were certain and true. In contrast, Stevin, Hariot, Cavalieri, Wallis, and their colleagues presented themselves as enterprising explorers of the mathematical landscape. Mathematics for them was a mysterious undiscovered land, which promised precious gems and marvels to the mathematician who would penetrate to its hidden recesses. The ultimate goal of a mathematician was not to deduce necessary truths, as it was for Clavius. It was, rather, to peer into the inner sanctums of mathematics and retrieve its secret treasures and wonders.[53]

THE HIDDEN TROVES OF MATHEMATICS:
INDIVISIBLES AS THE MATHEMATICS OF EXPLORATION

What does it mean to be a mathematical explorer? The answer is far from obvious. It was relatively easy for experimental philosophers to claim that they were discovering the hidden secrets of nature in the same manner as the voyagers explored the unknown corners of the globe. Natural philosophy, after all, had the entire natural world as its object. Exploring nature meant uncovering the unknown objects within it and the laws that govern it. Joseph Glanvill, writing in 1661, could confidently promise "that there is an *America* of secrets, and unknown *Peru* of Nature, whose discovery would richly advance the [Arts and Professions]."[54] The natural world, for Glanvill and his fellows, was indeed an undiscovered country, which must be mapped through systematic observation and experimentation.

Mathematics, however, presented a different case. Unlike natural philosophy, its only subject matter was given in advance: magnitude for geometry and number for arithmetic. Its fundamental laws were also predetermined, in the form of strict logical deduction for geometry and the basic arithmetical operations in the case of arithmetic. It is not at all clear what remains to be "explored" in a field in which both the subject matter and the fundamental procedures are known in advance. What could possibly be considered a hidden "gem" or "marvel" in a wholly necessary and predictable mathematical field? What, in other words, would a mathematics of exploration look like?[55]

A possible answer is suggested by a letter from Cavalieri to Galileo, written after he had received a copy of the *Two New Sciences*. "I am overcome with amazement," Cavalieri writes, "seeing with what new and singular manner [your work] unfolds the most profound secrets of nature, and with what facility it solves the most difficult things." Quoting Horace, Cavalieri compares Galileo to "the first to dare to steer the immensity of the sea, and plunge into the ocean" and continues:

> It can be said that with the escort of the good geometry and thanks to the spirit of your supreme genius, you have managed easily to navigate the immense ocean of indivisibles, of vacuum, of light, and of a thousand other hard and distant things that could shipwreck anyone, even the greatest spirit. Oh how much the world is in your debt for having paved the road to things so new and so delicate! . . . and as for me, I will be not a little obliged to you, since the indivisibles of my *Geometry* will gain indivisible lustre from the nobility and clarity of your indivisibles.[56]

The basic outlines are, of course, familiar by now: the mathematician is presented as a heroic navigator, traversing immense and treacherous oceans in search of "fine and delicate things." For the first time, however, the method of mathematical discovery is also named: it is the method of indivisibles, that highly controversial and extremely useful approach that dominated seventeenth-century mathematics and led eventually to the Newtonian and Leibnizian Calculus.

Galileo's disciples reiterate their view of the method of indivisibles as the proper vehicle for the exploring mathematician in other places as well. When Torricelli congratulated Cavalieri for opening the "royal road" to mathematics, he was referring explicitly to his colleague's "geometry of indivisibles," which "allows the establishment of innumerable and almost impenetrable theorems by short, direct, and positive demonstrations."[57] In a similar vein, Cavalieri thanked his friend for sending him his unpublished results, calling them "precious fruit" that he liked so much that he "can say nothing but call lucky the indivisibles that have found such a great promoter." Although these "marvelous finds" will, in Cavalieri's opinion, be admired by all, he holds little hope that they will change the minds of those "who abhor the indivisibles" as long as "the difficulties of the infinite keep their minds cloudy and hesitant."[58]

The precious fruits and marvels of mathematics, it seems, should be approached through the method of indivisibles. In Cavalieri's letter to Galileo, the method guides the mathematician through the turbulent "ocean of indivisibles" and leads to the great riches beyond. In the exchange between Cavalieri and Torricelli it is the royal road through the mathematical roughs, which breaks through seemingly impenetrable obstacles and clouded minds to retrieve the hidden secrets of mathematics. The method of indivisibles, in other words, leads the explorer into the mathematical heartland and enables him to chart its wondrous terrain.

But why? What is it about the method of indivisibles that makes it particularly appealing to those who view mathematics as an adventure and a voyage of discovery? In addressing this question, let us look at a simple example of proof by indivisibles offered by Cavalieri. It should be noted that more than any other mathematician, Cavalieri was the one most closely associated with the method of indivisibles and its dissemination. His *Geometria indivisibilibus* of 1635 and the subsequent *Exercitationes geometricae sex* of 1647 were the most serious attempts to turn what was a rather loose collection of mathematical practices into a systematic "method." All subsequent discussions of indivisibles in the seventeenth century refer repeatedly to his work, which was the most systematic and comprehensive in the field.

In proposition 19 of the first book of the *Exercitationes*, Cavalieri sets out to prove the following: "If in a Parallelogram a diagonal is drawn, then the parallelogram is double either of the triangles which are constituted by this diagonal"[59] (see Figure 27). Stated in nongeometrical terms, the theorem states that if a parallelogram is divided by a diagonal, its two triangle-shaped halves are equal to each other. The theorem is, of course, trivial and could be easily proved by a traditional Euclidean approach. All one has to do is show that the two triangles ACF and CDF are congruent (that is, identical). This follows immediately from the fact that the diagonal FC is a side of both triangles, angle ACF = angle DFC because AC is parallel to FD, and angle FCD = angle AFC because AF is parallel to CD. As a result the triangles share a side and have two equal angles; therefore they are congruent. Since the two together compose the parallelogram $ACDF$, then it is obviously double either of them.

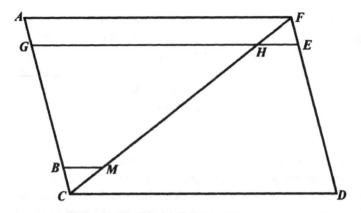

Figure 27. Bonaventura Cavalieri, diagram accompanying proposition 19 in *Exercitationes geometricae sex* (Bologna, 1647).

Cavalieri, however, proceeded differently. He divided each of the triangles into an infinite number of lines parallel to the bases *CD* and *AF*. These were the "indivisibles." He then argued that since each of the lines in one triangle had its equivalent in the other (for example, *HE* = *BM*), then "all of the lines" in one were equal to "all of the lines" in the other. From this he concluded that the areas of the two triangles were equal, and therefore the parallelogram is double either of them.[60]

The difference between the two approaches is striking: the Euclidean proof relied solely on rigorous deduction from first principles. Essentially, it showed that the parallelogram had to be double the triangles or a logical contradiction would ensue. It tells the reader nothing about the causes of the mathematical relationship, but simply shows that it cannot be otherwise. For Bacon, this would be a typical example of geometry remaining in the "open plains of generalities" instead of seeking a true cause in the "particulars." Cavalieri's strategy, by contrast, is to try to look into the inner structure of each triangle and determine its composition. His purpose is to "see" why the two triangles are indeed equivalent, not to logically prove that they could not be otherwise. By dividing the two triangles into their inner components, Cavalieri was exploring and mapping their inner structure, which was completely inaccessible as well as irrelevant to a traditional geometrical approach. Whereas the Euclidean proof relied on the strict application of the necessary rules of logic to the external characteristics of the parallelogram, Cavalieri's proof was an exploration of the internal composition of geometrical figures.

Cavalieri undoubtedly considered his system to be mathematically rigorous and strongly defended it against critics who pointed out inconsistencies in his method. Logical consistency, however, was not the structuring principle of his mathematical approach, as it was for Euclid. The aim of his method of indivisibles was to penetrate the surface appearances of geometrical objects and observe their actual internal makeup.[61] Francesco Stelluti, who was a contemporary of Cavalieri and an early member of the Accademia dei Lincei, claimed that the academy's purpose was to "penetrate into the inside of things in order to know their causes and the operations of nature that work internally."[62] Cavalieri, in effect, was doing precisely that for a geometrical figure: abandoning the traditional mathematical goal of logically proving his claim, he chose instead to "peer into" the parallelogram in order to "see" its inner workings. The Jesuit mathematician Paul Guldin, who was the leading critic of Cavalieri and his method, was undoubtedly correct when he charged Cavalieri with practicing a "geometry of the eye."[63]

While Cavalieri was perhaps the best-known "indivisiblist" of his time, Galileo and Torricelli were also closely involved in the development of infinitesimal methods and were well-known advocates of the approach. Galileo himself offered his own exploration of the inner structure of the geometrical continuum in the first day of his *Two New Sciences*. His discussion centers on several paradoxes, the most famous of which is known as "Aristotle's Wheel" (see Figure 28). When the large circle is rolled a full cycle along the line *BF*, every point on the wheel's circumference touches every point along *BF*. At the same time the smaller circle completes a full revolution along the line *CE*, touching the line at every point. The problem is that *CE*, being equal to the circumference of the larger circle, is longer than the circumference of the smaller circle. How is it possible for the smaller circle to produce a line longer than its own circumference through a single revolution, without dragging and while touching that line at every point?[64]

Galileo considers this a difficult problem and repeatedly refers to it as a "miracle" and a "wonder."[65] He does, nonetheless, offer a definite solution: the seemingly continuous line, he suggests, is in fact composed of an infinite number of points separated by an infinite number of empty spaces. In this way, the line *CE* is produced by the smaller circle through the interspersion of wider spaces between its indivisible points than the line *BF*, which was produced by the larger circle.

Galileo was well aware of the difficulties in this view and was careful to qualify his position. "What a sea we are gradually slipping into without knowing it!" his spokesman Salviati exclaims. "Among voids, infinites, indivisibles, and instantaneous movements, shall we ever be able to reach harbor even after a thousand discussions?"[66] His answer is a guarded "yes."

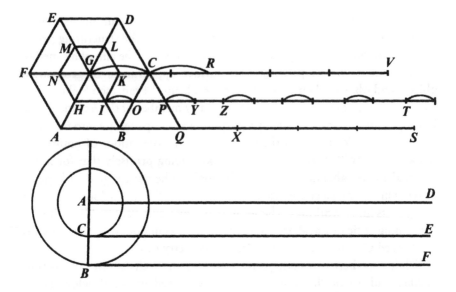

Figure 28.   Galileo Galilei, the paradox of Aristotle's wheel, from the *Two New Sciences.*

One should "allow this composition of the continuum out of absolutely indivisible atoms. Especially since this is a road that is perhaps more direct than any other in extricating ourselves from many intricate labyrinths."[67] By positing the continuum as composed of discrete indivisible components, its dangers and difficulties may be averted and the mathematician succeed in reaching "dry land."

Like Cavalieri, Galileo seeks to look into the invisible internal structure of a geometrical construct—the mathematical continuum. He is not content to merely register necessary geometrical connections in the Euclidean manner, but tries, instead, to determine the inner causes of the surface relationships we observe. His tool is the paradox: by pushing familiar geometrical relations to their incomprehensible limit, Galileo hopes to determine the structural causes that shape geometrical relations. Much as Bacon sought to vex and torture nature in order to extract her hidden secrets, so Galileo is intent on pushing geometry to its extremes in order to gain access to the miracles and wonders it withholds. In the absence of divine revelation, he argues, such "human caprices" are our best "guides through our obscure and dubious, or rather labyrinthine opinions."[68]

Torricelli, along with Cavalieri, was the leading "indivisiblist" mathe-

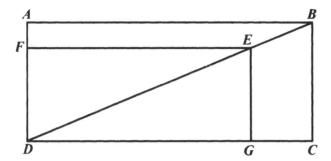

Figure 29.   Evangelista Torricelli's rectangle paradox,
in "De Indivisibilium Doctrina Perperam Usurpata."

matician in Italy and devoted far more time and effort to the development of
infinitesimal techniques than did his mentor, Galileo. Nevertheless, Galileo's
and Torricelli's mathematical approaches were markedly similar.[69] Classical
mathematicians had avoided paradoxes at all costs and sought to exclude
them from the field of legitimate mathematical investigations; in contrast,
both Galileo and Torricelli reveled in paradoxes and saw them as keys to true
mathematical understanding. Torricelli's papers include no less than three
separate lists of paradoxes of the continuum, starting with straightforward
ones and gradually increasing in difficulty and sophistication.[70] The most
basic one is simple, however, and already reveals Torricelli's fundamental
insight into what he considered the true nature of the continuum.

In a manuscript entitled "De Indivisibilium Doctrina Perperam Usurpata,"
Torricelli posits a rectangle, *ABCD*, which is dissected by the diagonal *BD*
(see Figure 29). From every point along *BD* lines are drawn parallel to *AB*
and *BC* respectively, until they intersect with the sides (like *FE* and *EG*). Now
the sum of all the lines *FE*, parallel to *AB*, compose the triangle *ABD*.
Similarly, the sum of all the lines *EG*, parallel to *BC*, compose the triangle
BCD. *FE*, however, is longer than *EG*, and this holds true for any point *E*
along the diagonal. The triangle *ABD*, composed of all the lines *FE*, must
therefore be larger than the triangle *BCD*, composed of an equal number of
shorter lines, *EG*. But that is manifestly false because the diagonal divides the
rectangle into two equal triangles.

How is it possible, Torricelli asks, that a collection of long lines produces
a figure of the same area as an identical number of shorter lines? The solution, he argues, is simple: "The opinion that indivisibles are equal among

themselves, that is that points equal points, lines are equal in magnitude to lines, and that surfaces are equal in depth to surfaces, is not merely difficult to prove, but is, in fact, false."[71] According to Torricelli, indivisibles have a positive magnitude. Some points are "larger" than others, some lines are "wider" than others, and some planes are "thicker" than other planes. In the case of the two triangles, he concludes, the indivisible lines EG, although shorter than the lines EF, are also "wider" than their counterparts. This explains how a collection of short lines produces a triangle of the same area as the same number of longer lines. It is, for Torricelli, the only possible solution to the paradox.[72]

Like Galileo, Torricelli is attempting to unveil the true structure of the continuum, which appears opaque and impenetrable at first sight. Galileo had examined the paradox of Aristotle's Wheel and argued that indivisible points are separated by empty spaces. Torricelli proposed his own paradox and concluded that indivisibles differ in magnitude from each other. Antonio Nardi, a disciple of Galileo and a close friend of Torricelli, once complained that while Archimedes arrived at marvelous mathematical results, his classical method provided "no trace of his voyage, no markers, and no guide."[73] Both Galileo and Torricelli were attempting to make up for Archimedes' shortcomings: they explored the very heart of the mathematical continuum and produced a detailed map of their discoveries for the benefit of those who would follow.[74]

Galileo, Cavalieri, and Torricelli knew each other well and were part of a tight-knit group of natural philosophers and mathematicians commonly known as the Galilean School.[75] It would only be expected that they should hold similar views on the nature of mathematics and the proper way to pursue it. Simon Stevin, however, belonged to a different world altogether. Born in Bruges in 1548 he was older than the Galileans. He was a practicing engineer who spent his life in the service of the Dutch States General and Prince Maurice of Nassau, supervising fortifications for the ongoing war with Spain.[76] Stevin often described the study of mathematics as a voyage of discovery aimed at unveiling secret hidden treasures; his mathematical approach well reflected these views.

A simple example can be found in his work on statics known as *The Art of Weighing*, published in Leyden in 1586.[77] In theorem 2 of volume 2 of this work, Stevin seeks to prove that "the center of gravity of any triangle is on the line drawn from the angle to the middle point of the side" (that is, on the median) (see Figure 30).

It should first be noted that even posing the question suggests a materialistic conception of geometrical figures. Euclid, after all, never attributed char-

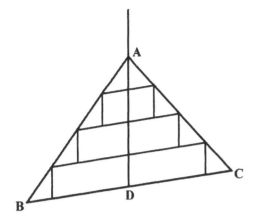

Figure 30.   Simon Stevin's depiction of a triangle's center of gravity.

acteristics such as "weight" and "gravity" to his circles and triangles, since they were purely immaterial Platonic constructions. Notably, Archimedes did treat geometrical figures as material objects, at least heuristically, but he always insisted on proving his results in a Euclidean deductive manner. Stevin, however, is unapologetic about his materialistic intuitions: by looking for the triangle's center of gravity Stevin indicated that he was treating it as a physical object and that he intended to locate a particular point "inside" it. This approach fits well with the style of mathematics practiced later by Cavalieri and Torricelli, which sought to look into geometrical figures and decipher their hidden secrets.

Stevin's method of proof further confirms his affinity with the mathematics of exploration. After drawing the median *AD* to the triangle *ABC*, he inscribes a series of parallelograms of equal height within the triangle (see Figure 30). The center of gravity of each parallelogram lies on the median because it divides each parallelogram into two equal parts, which are therefore in equilibrium. Consequently, the center of gravity of all parallelograms combined also lies on the median. Now it is possible, Stevin argues, to inscribe an infinite number of parallelograms inside the triangle, and their center of gravity will remain along the line *AD*. Furthermore, the greater the number of inscribed parallelograms the less their sum will differ from the area of the triangle itself. Now if we suppose that the triangles *ABD* and *ADC* are not in equilibrium, then there should be a fixed difference between them. But there can be no such difference, since each of them can be made to differ by less than any assigned quantity from the parallelograms on its side, which are in equilibrium with those on the other side of *AD*. *ABD* and

ADC must therefore be in equilibrium, and the center of gravity of the original triangle *ABC* lies on the median.

Stevin's basic strategy, then, is quite similar to Cavalieri's years later. He divides the triangle into what he considers its basic components and investigates its internal characteristics. The reason why the center of gravity lies on the median, he concludes, is that the triangle can be viewed as composed of an infinite number of parallelograms whose center of gravity is on the line. Stevin compared the mathematician to a mariner who brings back riches from his travels to exotic islands. His mathematical practice lives up to this imagery: he explores the hidden internal structure of geometrical figures and brings to light their hidden secrets.

In "weighing" geometrical figures and dividing them into their inner components, Stevin was following in the footsteps of Archimedes. Indeed, this ancient master was the model that Stevin explicitly sought to emulate in his work. Archimedes, however, was notorious among early modern mathematicians for deliberately obscuring his materialist and infinitesimal methods by using Euclidean deductive proofs in his formal presentations. Although these irrefutably demonstrated the result, they left no trace of the method of discovery that led to it in the first place.[78] Stevin, in contrast, had no qualms about basing his proofs on materialist intuitions and infinitesimal techniques and never bothered with rigorous deductive proofs. He sought to revive and perhaps improve on the methods of mathematical discovery, which the ancient master had sought to erase. In this he was a true adherent of the exploration approach to mathematics.

Although seventeenth-century English mathematicians did not embark on geographical voyages like their heroic Elizabethan predecessors, the imagery of exploration remained alive in mathematical circles. In Chapter 3 we noted William Oughtred's use of the imagery of exploration to describe mathematical studies. Oughtred, however, was not an adventurer: he was a tutor in a succession of noble households and probably never left England in his life. Despite this, he was imbued in the adventurous rhetoric of exploration mathematics. For him, as for his colleagues both in England and on the Continent, mathematics was concerned with exploring uncharted territories, unveiling their hidden secrets, and bringing them to light.

Perhaps because he viewed mathematics from the perspective of a teacher rather than that of an innovator, Oughtred differed in significant ways from the mathematicians already discussed. Many of his colleagues viewed the language of exploration as a promise of new and unsuspected knowledge awaiting mathematical discovery. They therefore concentrated their efforts on charting the inner composition of geometrical figures and bringing their secrets to light. Oughtred, as a teacher, was primarily concerned with elucidating existing mathematical knowledge for the benefit of his students, rather than pushing the boundaries of truth. It was therefore very much in character that for him, the labyrinth of mathematics lay not in the paradoxical continuum, but in the writings of the ancients. After all, "Euclides, Archimedes, Apollonius Pergaeus, that great geometer Diophantus, Ptolemaeus, and the rest" were the authors of the texts that were taught to all students of mathematics at the time.[79]

To Oughtred, these classical mathematical texts presented an interconnected web of logical relationships between fascinating and surprising results. One could follow the proofs and reasoning that led from one result to the next, but one could never truly capture the fundamental reason why they were true. For Oughtred, the ancient sources concealed as much as they revealed and left their reader puzzled and confused. The precious secrets of geometry remained hidden in the classical texts behind a mystifying and confusing network of logical deductions.

Oughtred sought to break through this bewildering web. His mission was to hand the "ingenious lovers of these sciences, as it were, *Ariadne*'s thread, to guide them through the intricate labyrinth of these studies."[80] Instead of following the twists and turns of the ancients' texts, he wants to provide an unobstructed view of mathematical truths: "Wherefore that I might more clearly behold the things themselves, I uncasing the propositions and demonstrations out of their covert words, designed them in notes and species appearing to the very eye."[81] The challenge of mathematics, for Oughtred, is to bring to light the hidden secrets of mathematics and present them clearly to the "eye."[82] This was, of course, precisely the view that characterized exploration mathematics.

Whereas Hariot, Stevin, and the Galileans attempted to provide detailed charts of geometrical surfaces and objects, Oughtred proposed different means to bring to light mathematical truths. His most significant contributions to mathematics lay in his introduction and diffusion of standardized algebraic symbols. His allusion to "notes and species" in the quoted passage refers precisely to his work in this area. In Oughtred's view, the ancient texts were opaque in large measure because they used plain descriptive text instead of algebraic symbols. Proper mathematical signs, he believed, would expose the true nature of the seemingly impenetrable web of mathematical reasoning. For Hariot, Stevin, and their Italian contemporaries, the rhetoric of discovery pointed the way toward a study of infinitesimals and the method of indivisibles. For Oughtred, it suggested that proper algebraic symbols are the proper tools to explore the hidden mysteries of mathematics.[83]

But although Oughtred was not engaged in the study of infinitesimal methods, he was nonetheless quick to appreciate their value. When late in life he was introduced to Cavalieri's new methods, he immediately recognized their importance for the mathematics of exploration. Around 1645 he wrote Robert Keylway that he was "induced to a better confidence of your performance by reason of a geometric-analytical art or practice found out by one Cavalieri, an Italian . . . [from which] I divine great enlargement of the bounds of the mathematical empire will ensue."[84] The method of indivisibles, Oughtred believed, offered great new vistas for new explorers to conquer. Unfortunately, he wrote Keylway, he was too old and dispirited by the civil war raging around him to embark on the voyage himself: "Being more stept in years, daunted and broken by the sufferings of these disastrous times I must content myself to stay home and not put out to any foreign discoveries."[85]

The method of indivisibles, for Oughtred, offered new possibilities for exploring the great mathematical unknown. Like the new mathematical symbols he advocated, it would be used to penetrate the inner recesses of mathematics, "uncase" its secrets, and present them "to the very eye."[86]

Although Oughtred declined the opportunity to embark on this grand adventure, he did encourage others to undertake the mission. When in 1655 John Wallis dedicated his *Arithmetica Infinitorum* to him, Oughtred responded by congratulating Wallis "both for such great clarity, as for the penetration of your mind and genius. Which will not only enter a new place but will open a way for others to the most hidden mysteries of the art of penetration." "Your invention then," he added, "full of mysteries, arouses me more than the somewhat small chapter which Sir Charles Cavendish showed me twenty years ago of an excellent new theorem devised by Cavalieri's method."[87] Wallis, according to Oughtred, was expanding the work of Cavalieri by penetrating even further into a "new place," bringing to light hidden mysteries and opening the way for others to follow. He was, in fact, embarking on the hazardous voyage that Oughtred himself declined to undertake.

Wallis shared Oughtred's enthusiasm for Cavalieri's work and viewed his own work as an extension and improvement of Cavalieri's method.[88] But whereas Cavalieri and his fellow Galileans remained close to their geometrical roots and Oughtred stressed the importance of algebraic notation, Wallis combined the two approaches. In his work, he brought the superior power of the new algebra to bear on the method of indivisibles, thus joining the two modes of exploration mathematics. Indeed, Wallis's great accomplishment, and the basis for his reputation, was his "translation" of Cavalieri's geometrically based "method of indivisibles" into an algebraic "arithmetic of infinites."[89]

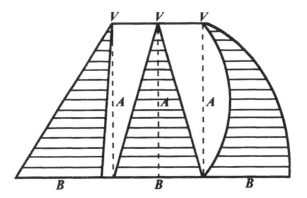

Figure 31. John Wallis, "The composition of geometrical figures." Graphic accompanying proposition III of *De sectionibus conicis*. Printed in John Wallis, *Opera mathematica*, vol. 1 (Oxford, 1699), 299.

A typical example of his mathematical practice can be found in the beginning of his 1655 tract *De sectionibus conicis*, where Wallis attempted to prove the elementary formula for calculating the area of a given triangle. His formulation of the problem and the manner in which he went about solving it reveal much about his mathematical approach. "Proposition III: Since a triangle consists of an infinite number of arithmetically proportional lines or parallelograms, from their beginning in a point continuously to the base, the area of the triangle will be equal to the base multiplied by half the altitude."[90] This is a highly unorthodox statement of the theorem. Wallis here simply assumes the highly questionable view that the surface of a triangle is somehow composed of an infinite number of infinitesimal lines or parallelograms. But although problematic from a traditional standpoint, the statement of the theorem is reminiscent of Cavalieri's (far more cautious) approach of dividing up geometrical figures into their fundamental components.

Wallis's proof is even more mathematically reckless than his initial formulation. He first calculates the sum of all the lines (or infinitesimally small parallelograms) that "compose" the triangle by treating them as an arithmetical progression. Since the first term in the progression, at its apex, is 0 (or o, a cipher, for Wallis), the last term, at the base, is B, and the number of terms is ∞, the sum of the progression is $B\infty/2$. He then multiplies this sum by the "height" of each line or parallelogram, which he calculates as the altitude divided by infinity, or A/∞. Since the two infinite terms cancel each other out, Wallis concludes that the area of the triangle is equal to $AB/2$ (see Figure 31).

Many of Wallis's contemporaries, most notably his French colleague Pierre de Fermat, were aghast at Wallis's uncritical manipulation of infinities and zeros.[91] What interests us, however, is not the technical merit of Wallis's proof, but rather the vision of mathematics that it encodes. Like Cavalieri, Wallis was not content to simply prove his result in the traditional Euclidean manner. Instead, he sought to peer into the hidden inner structure of the triangle and see how its area is composed. In order to calculate the triangle's area, he essentially added up the infinite number of fundamental components that comprise it and ultimately arrived at the familiar formula.

In his dedication to Boyle of *De Cycloide* Wallis expressed his view that mathematical analysis, which for him included infinitesimal methods as well as algebra, should be used to "search everywhere for the hidden, or extricate what is veiled."[92] In this simple example his meaning becomes clear: the hidden and opaque inner structure of a geometrical figure is unveiled, mapped out, and brought to light. Wallis is clearly not a classical mathematician, who wants to provide an irrefutable logical demonstration of his result. He is, rather, a mathematical explorer, who wants to penetrate through the opaque mathematical surface and chart its hidden secrets within. Oughtred had predicted that great new mathematical vistas would emerge from the method of indivisibles. Wallis set out to explore them.

Wallis's brand of mathematics was soon eclipsed by the very different approaches of Newton and Leibniz. His, however, was not the last effort at formulating an experimental approach to infinitesimals. In 1727, Bernard de Fontenelle, permanent secretary to the Paris Academy of Sciences, published a book in the philosophy of mathematics entitled *Elements de la géométrie de l'infini*. In the introduction he wrote:

> The calculus is to mathematics no more than what experiment is to physics, and all the truths produced solely by the calculus can be treated as truths of experiment. The sciences must proceed to first causes, above all mathematics where one cannot assume, as in physics, principles that are unknown to us. For there is in mathematics, so to speak, only what we have placed there . . . If, however, mathematics always has some essential obscurity that one cannot dissipate, it will lie uniquely, I think, in the direction of the infinite; it is in that direction that mathematics touches on physics, on the innermost nature of bodies about which we know little.[93]

Mathematics, according to Fontenelle, is much like physics when it touches on the infinite. For it is only in this field that, like physics, it seems to hold mysteries that we have not "placed there." These are the secrets of "the innermost nature of bodies," and the only way to elucidate them is through the "experiments" of the calculus. Fontenelle was heir to a long tradition that equated infinitesimal methods with an empirical search for the hidden structure of bodies. It began more than a century and a half earlier with Stevin and Hariot; its echoes were still heard in the eighteenth century.

Conclusion

In the year 1539, Georg Joachim Rheticus, professor of mathematics at the University of Wittenberg, set out to the Prussian town of Frauenburg to visit the reclusive astronomer Nicholas Copernicus. Rheticus had heard rumors of Copernicus's radical heliocentric cosmology and was determined to investigate its truth and validity firsthand. Under the old canon's tutelage, Rheticus quickly became an ardent Copernican. He ended up staying in Frauenburg for two years, often impressing on his host the significance of his discoveries and the importance of making them available to a wider audience. A few months after his arrival he composed a short treatise, summarizing the main points of the Copernican theory, and sent it to his colleague Johannes Schöner. It was published in 1540, and became known as the *Narratio Prima*, the first published account of Copernicus's views.[1] At the end of the treatise, Rheticus reflected on the hospitality offered him by narrating the adventures of the sophist philosopher Aristippus:

> The story is frequently told of the shipwrecked Aristippus, which they say occurred off the island of Rhodes. Upon being washed ashore, he noticed certain geometrical diagrams on the beach; exclaiming that he saw the traces of men, he bade his companions be of good cheer. And his belief did not play him false. For through his great learning he easily obtained for himself and his comrades from educated and humane men the things necessary for sustaining life.[2]

Rheticus goes on to report that among Prussians one can see "geometrical diagrams at every threshold" and "geometry present their minds." The kindness and generosity of his hosts, it seemed to Rheticus, was intimately related to their love of geometry.

For Rheticus as for Aristippus, geometry was the sign of human habitation, civilization, and safety. Aristippus was tossed about by stormy seas, until shipwrecked on an unfamiliar coast. Only when he saw geometrical figures in the sand did he realize that he had found the safety and security of civilized habitation. Two thousand years later, Rheticus undertook a toilsome journey from Wittenberg to Frauenburg, traversing the breadth of German-speaking lands. The geometrical diagrams he recognized "at every threshold" in Prussia signaled to him the end of his difficult journey. Now he could at long last enjoy the safety and hospitality of generous and learned friends, skilled in geometrical reasoning. From Aristippus to Rheticus, geometry signaled safety, stability, and security. It was the safe haven at the end of a journey, a refuge from the ravages of the sea.

The significance of Aristippus' geometrical refuge becomes clearer if we consider it as a metaphor on the nature of knowledge. When viewed in this light, the association of Euclidean geometry with safety and stability seems natural. Of all fields of ancient and medieval knowledge, geometry stood in the highest esteem for the certainty and universality of its results. The truth of a geometrical theorem, based on self-evident assumptions and proceeding by indisputable logical deduction, could not be challenged. Whereas the doctrines of philosophers, physicians, and even theologians all seemed to invite debate and conflict, Euclidean geometry stood alone, unchallenged, as a model of true and universal knowledge. It was indeed a haven of safety and stability in a hazardous and confusing ocean of knowledge.

But what had held true for more than two millennia was shortly to be challenged. When in 1614 John Davies of Hereford sought to praise the achievements of the mathematician Edward Wright, he did so in very different terms:

> Wright (ship-wright? no; ship-right, or righter then,
> when wrong she goes) lo thus, with ease, will make
> Thy Rules to make the ship run rightly, when
> She thwarts the *Maine for Praise or profits* sake.[3]

Not a trace remains here of the safety and civility of the geometrical shore. The mathematician (or "geometer") is here nothing but a privateer, braving hazardous oceans in search of glory and profits.

Hereford was not alone in promoting this new image of the mathematical sciences. Stevin in the Netherlands, Wright, Hariot, Oughtred, and their colleagues in England, and the Galilean circle in Italy all shared in the vision of mathematical study as an adventurous voyage of discovery. The mathematical sciences, which had traditionally signified certainty and stability, were now likened to a grand adventure upon the seas, whose potential gain is matched by the ever-present danger. In this book I have sought to trace this profound transformation in the way mathematics was described and perceived, track its origins, and outline its implications for mathematical practice. It has been the main thesis of this book that the imagery of a mathematics of adventure and exploration went hand in hand with the emergence of infinitesimal methods.

Early modern mathematicians were not the first to adopt the imagery of exploration to describe their intellectual pursuits. It was, rather, the experimental philosophers of the age who proposed a new program of research inspired by the geographical voyages. The new philosophers sought to move beyond the limits set by the established canon and ancient authority. Through systematic experimentation and firsthand observations of nature, they promised, new and wondrous knowledge will be gained that far surpasses the familiar truths contained in the traditional sources. As their models, the reformers adopted the great voyagers of the age, such as Columbus and Magellan. They were the first, it was said, who dared to break through the confines of the known universe and bring back the wonders and marvels of a new world. The new experimenters, following in their footsteps, will explore and chart the hidden realms of the natural world. The imagery of exploration thus proved to be ideally suited for the promotion of the experimental philosophy, which sought to explore the intricate pathways of nature and bring to light secrets and wonders never seen before.[4]

But while the imagery of exploration seemed a natural model for the practices of experiment and observation, such was not the case for the study of mathematics. Classical mathematics, and most prominently geometry, had long been admired for its clarity, the logical necessity of its deductions, and the incontestability of its claims. Its truths, although amazing and surprising to a novice, were fixed and familiar to the experienced practitioner and had not been challenged for almost two millennia. They were universal, irrefutable, and true. Mathematics, as practiced, had no place for the exploration of new and unknown terrain, or for the discovery of secret hidden marvels. In a system where all truths are implicit in the initial assumptions, one cannot hope to discover hidden wonders.[5]

Seemingly in response to this critique, a growing number of mathematicians across Europe sought to redefine their discipline and began presenting themselves in terms similar to the experimental philosophers. They fashioned themselves as enterprising explorers, seeking treasures in the uncharted oceans and lands of mathematics. Although acknowledging their debt to the classical tradition, they also emphasized its limitations. For them, traditional mathematics merely established necessary logical relations between its objects, but never penetrated to their inner core. It could describe certain inescapable truths about the world, but it could not show the fundamental reasons why these relations were true. A new approach was needed if mathematics was to break through the confines of its past and discover the treasures and marvels hidden within its fold, namely, a mathematics of exploration.[6]

The new rhetoric of exploration expressed itself in several ways in mathematical practice. Oughtred, for example, emphasized the value of proper mathematical symbols, which would present to the reader the previously hidden inner workings of mathematical proofs and arguments. For Wallis, "the way of . . . finding out of things yet unknown (which the ancients did studiously conceal)" was, on occasion, his inimitable "method of induction."[7] Even Clavius, who in the 1580s insisted on the strict geometrical and deductive nature of mathematics, in 1609 used the trope of exploration to describe the purpose and meaning of algebra.[8] Most importantly, however, the new rhetoric of exploration was implicated with the investigation of the inner and hidden structure of geometrical figures. This study was the essence of the method of indivisibles.

It is easy to see why the method of indivisibles proved so appealing to mathematicians who viewed themselves as explorers and discoverers. The method was designed precisely as an exploration of the inner composition of geometrical figures. Indivisiblist mathematicians began their investigation by "peering into" a geometrical figure and determining its internal makeup. Some, like Galileo and Torricelli, sought to further explore the nature and composition of the geometrical continuum by using paradoxes to test the limits of mathematical reasoning. Others, like Cavalieri and Wallis, investigated the true inner causes of familiar results such as the area of a parallelogram or a triangle. Either way they were aggressively mapping out the inner structure of geometrical figures in ways that traditional geometry never attempted. It may be said that traditional geometry sought to determine the nature of the world through rigorous deduction from first principles. In contrast, Cavalieri and his colleagues sought to investigate the world by exploring the secrets hidden within existing objects. The method of indivisibles was indeed a voyage of discovery into the hidden secrets of mathematics.

This was a far cry from the safety of Aristippus' geometrical haven. With varying degrees of enthusiasm, Hariot, Cavalieri, and their colleagues challenged the old certainties and introduced far more problematic, paradoxical, and controversial mathematics. Rather than deducing certain conclusions from self-evident assumptions, they sought to explore and map out the hidden internal structure of geometrical figures. In doing so, they were willing to openly confront paradoxes and uncertainties and sometimes even welcome them into the fold of their geometrical method. The safe haven of Euclidean geometry was left far behind. The new mathematics was an adventure, a risky voyage on stormy seas, whose promise of hidden gems was matched by the danger of shipwreck on the shoals of paradox and uncertainty. In the eyes of its practitioners, it was a hazardous adventure whose ultimate end could only be guessed at.

Before concluding, I would like to add a few words on the subsequent fortunes of the mathematics of exploration. John Wallis's work may well have signaled the end of this mathematical tradition that began almost a century before. His combination of an infinitesimalist approach with an algebraic exposition is often cited as leading directly to the calculus of Newton and Leibniz. In this sense one can conclude that the calculus was the culmination of exploration mathematics. With the calculus, as Oughtred had hoped, the study of infinitesimals opened up new vistas of understanding of the mathematical as well as the physical world.

At the same time, it must also be acknowledged that the calculus signaled a clear break from the mathematics of exploration. Unlike the infinitesimal methods that preceded it, the calculus was first and foremost a self-contained and internally consistent mathematical system. It was completely independent of geometrical figures and did not seek to "penetrate" their surface and chart their inner structure. In fact, it had no need for external objects at all: its fundamental theorem stated an inverse relationship between the two algebraic operations of differentiation and integration (to use Leibnizian terminology). It was a self-contained system that need only be logically coherent and internally consistent. And although it could be applied to a myriad of objects and processes and used to describe them, the truth of its propositions was not dependent on any of these. It was merely a function of its own internal consistency. In this sense, the calculus had more in common with traditional Euclidean geometry than with the infinitesimal methods that preceded it.[9]

What began with Hariot and Stevin as a radical departure from the self-contained Euclidean deductive system, ended with Newton and Leibniz as a self-contained and internally coherent algebraic calculus. The mathematics of exploration had come full circle.

Reference Matter

Appendix A

THE MATHEMATICAL NARRATIVE

Scientific practice is above all a story-telling practice.

— DONNA HARAWAY, *Primate Visions*

The history of mathematics has always been a problematic field in the larger discipline of historical study. History to its practitioners is the study of particulars: specific periods and events are studied for their own sake, as unique occurrences that took place within human history. Mathematics, of course, lies at the opposite pole of the intellectual spectrum. In stark contrast to historians, mathematicians claim to study universal and necessary truths. Geometrical proofs and algebraic calculations are founded on self-evident assumptions and are elaborated systematically in accordance with unchanging universal laws. They are expected to hold true under any circumstances, in any time or place, quite independently of the vicissitudes of human society. If history is the study of the contingent, then mathematics, as commonly understood, is concerned with Platonic universals.

It is hardly surprising then that the combination of such opposing intellectual traditions has often produced an uneasy liaison. How can one study the peculiarities and contingencies of a field dedicated to the formulation of universal, unchanging truths? Can mathematics, in other words, truly have a history? Traditional works in the history of mathematics played down the contingent aspects of historical works in favor of the transcendence of the mathematical subject matter. Mathematical innovations are described as unfolding progressively, moving from the simple to the complex in accordance with the subject's inescapable inner logic. Mathematics, in such works, unfolds *within* history—various men (for in such works there is hardly a woman to be found) in different locations and time periods discover various truths that contribute to the totality of mathematical knowledge. Mathematics is not, however, *of* history, for the content and the validity of its theorems are completely independent of the particular social setting that spawned them. In traditional history of mathematics, in other words, human history is the theater of mathematics, but not its dimension.[1]

The traditional form of the history of mathematics accords well with professional mathematicians' understanding of their discipline as pure and abstract. Scrupulously adhering to the demands of the mathematical ethos, such works observe a strict separation between the "mathematical" and the "social" components of their narrative. The former constantly moves forward, powered by its own irrepressible logic, while the latter is reduced to a collection of historical anecdotes drawn from the lives of practitioners of mathematics. Far from being an attempt to imbricate mathematical knowledge with its temporal setting, these anecdotes serve an opposite purpose altogether: the abstract beauty and orderliness of mathematics is brought into clear relief when contrasted with the untidy and seemingly senseless collection of facts that comprises human history.

It was only to be expected, then, that historians and sociologists over the years have found this form of the history of mathematics to be unsatisfactory. Indeed, a form of study that emphasizes the progressive unveiling of universal truths, rather than the contingencies of human existence, has little to do with "history" as it is commonly understood. Writing a different kind of history of mathematics, however, was a difficult proposition, as historians and sociologists were quick to note. The essential problem was best expressed by Karl Mannheim, who claimed that "even a god could not formulate a proposition on historical subjects like 2×2=4."[2] He then proceeded to exempt the study of mathematics from his sociology of knowledge.

Mannheim was undoubtedly representative of most historians and sociologists, who avoided mathematics and devoted their energies to more promising fields of study. Some, however, took on the challenge and confronted the problem directly. Although approaching the issue from a wide variety of political and methodological starting points, they shared a common goal of providing a social and historical account of mathematics.[3]

An early and intriguing effort in this direction was made by Oswald Spengler, who provided one of the boldest cultural interpretations of mathematics to date. Spengler was famous for his mystical approach to the history of civilizations and viewed mathematics as the actualization of a culture's picture of the natural world. The history of mathematics was for him the ultimate test for his claim that all aspects of a given civilization were connected by a unifying cultural principle. "Between the Differential Calculus and the dynastic principle in politics, between the Classical city-state and the Euclidean geometry, between space perspectives of Western oil-painting and the conquest of space by railroad, telephone, and long-range weapon . . . there are deep uniformities," he wrote in the introduction to The Decline of the West.[4] If "pure" mathematics could be shown to be dependent on its cultural setting, Spengler reasoned, then surely the less abstract fields of human activity could be shown to conform to underlying cultural principles. True to his word, Spengler began his opus with a chapter entitled "The Meaning of Numbers," where he argued that the notion of number is unique to any given culture. "*There is not, and cannot be, number as such.* There are several number-worlds as there are several Cultures," he concluded.[5]

At the opposite end of the political spectrum, Marxist sociologists were also attempting to incorporate mathematics into a larger historical framework. Whereas Spengler interpreted mathematics as a reflection of the "Soul" of a civilization, for

Marxists it was necessarily dependent on the material conditions of production within a given society. Boris Hessen, in his classic 1931 paper "The Social and Economic Roots of Newton's *Principia*" argued for an economic basis for Newtonian mathematical physics, but stopped short of claiming that the actual content of mathematics was dependent on social conditions.[6] Franz Borkenau took a more radical position, arguing that the very nature of mathematical abstraction is the result of the growing division of labor, where action becomes quantifiable and particular qualities are lost.[7] Richard Hadden has recently revived Borkenau's thesis in his *On the Shoulders of Merchants*, but with important differences: where Borkenau emphasized the production process as the origin of mathematical abstraction, Hadden argues that the true source is the capitalist mode of exchange, which reduces all commodities to uniform, numerical "values."[8]

More recently, mathematics became a test case for the social approach to the study of science known as the Strong Program. David Bloor viewed mathematical knowledge as the greatest challenge to the Strong Program's sociological interpretation of scientific knowledge. In this he was following in the footsteps of Spengler and his Marxist rivals, each of whom viewed mathematics as the most challenging case to their respective systems. Clearly, if mathematics could be shown to be a form of social knowledge, then other, less abstract fields could certainly be interpreted as social through and through. Bloor does not so much offer a social history of mathematics, as argue for its possibility in principle. By drawing on the insights of John Stuart Mill and Ludwig Wittgenstein, Bloor offers a vision of mathematics as based ultimately on material experience, but shaped by social convention.[9]

In the past decade Sal Restivo has offered his own sociological reconstruction of the history of mathematics. According to Restivo, purity and abstraction should be understood as the effects of the social process of the professionalization of mathematics. Each generation of mathematicians, according to Restivo, hands down to its successor a set of theories and results. These then become the raw material for the construction of a new level of mathematical investigations, whose results will be handed down to the next generation in turn. This process, according to Restivo, produces an ever-greater degree of abstraction with each passing generation. The increasing professionalization of mathematics guarantees the continuity of this self-perpetuating process, as well as a relative immunity to external factors that might interfere with its smooth operation.[10]

Overall, the would-be historicizers of mathematics utilize two main strategies.[11] One approach, popular with Marxists as well as with Bloor's Strong Program, strives to demonstrate that mathematics is based on perceptions of matter and on social practices. Bloor, for example, uses pebble games developed by Z. P. Dienes in order to demonstrate the material basis of an equation like $(x+2)x+1 = (x+1)^2$.[12] Similarly, Hadden suggests that the mathematical concept of "general magnitude" originated in commercial practices of exchange at the dawn of the capitalist age. The purpose of this approach is, of course, to poke holes in the self-image of mathematics as universal, pure, and abstract. If mathematics can be shown to have a concrete physical and social basis, then it can also be shown to be an integral part of human historical civilizations.

The alternative approach accepts the status of mathematics as pure and abstract, but attempts to provide a sociological explanation for this state of affairs. Restivo's views on mathematical abstraction fall into this category. Mathematics itself, according to Restivo, can indeed be understood as an abstract logical structure; these characteristics, however, are the result of the social and historical process of professionalization. A more extreme example of this style of argument can be found in Philip Kitcher's *The Nature of Mathematical Knowledge*.[13] In Kitcher's view, mathematics originated in the material experience of our ancestors, most likely in ancient Mesopotamia. Since then, however, the knowledge has been preserved and developed through a social process—the unbroken passage of mathematical knowledge from teacher to student in each succeeding generation.

Restivo's account of abstraction and Kitcher's views on mathematical knowledge are probably useful for a philosopher attempting to provide a non-Platonist account of mathematical truths. Both Restivo and Kitcher, after all, treat the development of mathematics as a social matter, and Kitcher even presupposes that the ultimate source of this knowledge is material. If our sole interest is in the philosophical question of whether mathematics is perfectly pure and abstract, then Restivo and Kitcher answer with a well-founded and convincing "no." If, however, we are interested in the actual historical relations between mathematics and its social and cultural setting in a given period, then their approach would be of little help. By focusing on the preservation of mathematical purity through its containment in a single great tradition, Restivo and Kitcher effectively recreate the boundaries between mathematics and society that they sought to bring down. In the traditional view, mathematics was insulated from society because it dealt with ahistorical Platonic universals. In Restivo and Kitcher's model, mathematics is just as effectively insulated from wider cultural trends; the only difference is that the insulation is caused by certain social factors, rather than by the essential nature of mathematics. All this, of course, does not imply that Restivo and Kitcher are necessarily wrong in their analysis. It simply suggests that such an approach will not result in embedding mathematical practices and techniques within their historical culture.[14]

The Marxists and the adherents of the Strong Program appear at first glance to offer a bolder and more promising approach. Their attempts to imbricate mathematics with historically placed social and economic factors are precisely the kind of studies needed in order to truly historicize mathematics. My sole criticism of their efforts is that their project appears to be overwhelmingly difficult. In essence, they are launching a frontal assault on the self-understanding of mathematics at its very core.

Mathematicians view their subject as logical, rigorous, abstract, and self-contained. Whatever the precise meaning attached to those terms, it must be admitted that those depictions are most often accurate accounts of the discipline. Whenever Marxists or Strong Programmers attempt to place mathematics within a historical context, they must first dismantle this fundamental account of mathematics. They must demonstrate that the elaborate apparatus protecting the "purity" of mathematics does not apply in the particular historical setting on which they focus. I am certainly not claiming that adherents of this approach are always unsuccessful in their attempts to establish the historicity of mathematical practices and techniques. My point is, rather, that a methodology that necessitates an attack on the abstract nature of mathematics whenever it needs to make a historical point, must, in the end, be of limited use.

It is precisely for this reason that I find a narrative approach to be a most promising avenue for historicizing mathematics. Rather than chipping away at the abstractness and rigor of mathematics, this approach makes use of precisely these qualities in establishing the cultural dependency of mathematical constructs. For while it is always difficult to point to the presence of social ideas and practices within mathematics, mathematical work does, I argue, contain a narrative. Once this narrative is identified, it can be related to other, nonmathematical cultural tales that are prevalent within the mathematicians' social circles. A clear connection between the "mathematical" and the "external" stories would place a mathematical work firmly within its historical setting.

The conceptual approach to mathematics that I am relying on is known as "formalism" and is associated with the work of David Hilbert in the first quarter of the twentieth century.[15] It must be admitted that using Hilbert's approach in a work dedicated to establishing the historical context of mathematical practices may seem like a curious choice to historians and philosophers of mathematics. After all, Hilbert viewed mathematics as pure, abstract, and essentially ahistorical, a universal and rational language that completely transcends its local roots. He would be very much opposed to any attempt to anchor mathematics to cultural practices and mores and would therefore seem to be an ill-suited candidate for the project of this book.

It is, however, precisely Hilbert's insistence on the purity and internal coherence of mathematics that I find most useful for my project. My purpose is to show that one can accept the essentials of Hilbert's formalist view of mathematics without compromising a cultural and historical approach to the field. Mathematics can be everything that Hilbert claims that it is—pure, rigorous, abstract, and self-contained—and yet be culturally grounded in a particular historical reality. This is accomplished by means of the "mathematical narrative," which bridges the divide between the abstract and social views of mathematical practice.

According to Hilbert, any mathematical construction must begin with a set of "axioms" or "postulates." These must be stated before it is known whether any objects exist that fulfill their conditions. The consequences of the postulates are then elaborated, and the theory produced is presumed to apply to any system of objects (if such exists) for which the axioms hold true. From the formalist perspective, the mathematician in essence creates a self-contained world. He selects a series of assumptions from an unlimited number of possibilities and deduces truths that apply to objects fitting the postulates. He is constructing a mathematical universe that accords with his initial presuppositions and is ruled by the theorems that follow from them.

Geometry is the perfect example for this view of mathematics. From antiquity to the nineteenth century, Euclidean geometry was viewed as a set of incontrovertible truths. It was based on simple and undeniable assumptions about the world, known as "axioms." From these, Euclidean geometry deduced an impressive set of conclusions of seemingly universal applicability. In the nineteenth century, the work of Riemann, Bolyai, and Lobachevskii demonstrated that alternative, non-Euclidean geometries were also possible. By positing a different set of assumptions, one could construct a different, but consistent, non-Euclidean world. This "world" may not accord with our daily experience as well as Euclidean space, but it does, nonetheless, describe true relations of objects in a space that conforms to the system's axioms. From a formalist perspective, Euclidean geometry is in no way "truer" than the non-Euclidean varieties. Each is a true "world" as long as it accords with its own initial postulates.[16]

Hilbert and other formalists would insist that, in principle, a mathematician is not constructing a new world, but simply deducing necessary conclusions from a random set of propositions. Strictly speaking, the postulates are mere symbolic strings that have no meaning in and of themselves. By applying certain "operations" to them, the mathematician arrives at new propositions, which are also nothing but empty formulas. In actual mathematical practice, however, this is not the case. Mathematicians do not select their postulates at random, but according to preconceived notions of their meaning and significance. Geometry, for instance, whether Euclidean or not, can in principle be treated as a set of contentless formal propositions. In practice, of course, both Euclid and his nineteenth-century successors selected their axioms with great care and proceeded to deduce certain theorems in accordance with their prior notions of "space." Riemann, Bolyai, and Lobachevskii, like Euclid before them, were not playing an empty formula game. They were creating spatial universes, each obeying different geometrical laws.

The world-building enterprise implicit in this conception of mathematics was noted by Sal Restivo, who compared the formalist mathematician's work to that of a novelist: "In the formalization process," he wrote, "what happens in effect is that the mathematician explicitly creates a mathematical reality very much in the way a Tolkein or a Frank Herbert creates a fantasy world."[17] Restivo's point is well taken. The mathematician, like the novelist, creates his or her self-contained world. In both cases (despite the protestations of principled formalists), this universe is inspired in certain ways by our perceptions of "our" world, and in both cases it also diverges from perceived reality in certain important ways. Both novelist and mathematician begin by setting down the basic premises of their newly created "world." They then proceed to follow their development through the twists and turns of the plot.

It will likely be pointed out that mathematicians are constrained by much tougher restrictions than novelists when they venture to construct a hypothetical "world." A mathematical theory must be formally and precisely self-consistent and developed only through strict rules of reasoning. A work of fantasy, in contrast, follows only much looser standards, such as internal coherence and a degree of plausibility. Nonetheless, despite being more heavily restricted, the mathematician is free to make many of the same choices as a fiction writer in producing his own universe. The mathematician determines the premises for his construct and then chooses in which direction to develop them in order to produce the system he or she desires. In many historical periods he may also grant himself an impressive degree of latitude in applying the rigorous standards of mathematics.[18] In a work of fiction, those very elements—the premises of the story, its development, and even the type of inner logic it follows—can be recognized as the work's narrative structure. My claim is that such a narrative can be detected in mathematical works as well.

Once a particular narrative has been located in a mathematical work, it can be used to establish the fundamental relationship between the mathematical practice and its specific historical setting. The mathematical tale must simply be compared to other, nonmathematical tales, which were prevalent in the wider social and cultural circles in which the mathematicians worked. If a strong relationship can be established between a historically specific nonmathematical tale and the narrative of a mathematical work that originated within its social sphere, then mathematics can indeed be said to be fundamentally shaped by its social and cultural setting.[19]

The methodological advantages of a narrative approach to historicizing mathematics are clear. Marxist and Strong Program efforts had to argue forcefully that mathematics is not really as formal and abstract as it appears. Although such a strategy appeared necessary in order to break down the conceptual barriers between mathematics and its historical setting, it was also risky and difficult to maintain. In some instances it appeared that such projects attempted to redefine mathematics rather than explain existing mathematical practice. A narrative approach, by contrast, not only accepts mathematics as formal and abstract, but insists on those very characterizations. It is precisely the internal coherence of mathematics and its clear formal structure that mark it as a good candidate for a narrative interpretation. The abstract and formal nature of the discipline, which previously appeared as an obstacle to historicizing mathematics, has now become a major tool for placing mathematics firmly within its cultural setting.[20]

This book is a narrative-based interpretation of an episode in the history of mathematics and is intended as both a methodological work and a historical contribution. On a methodological level, the book is a demonstration of a narrative approach to the history of mathematics. By using this methodology, complex mathematical techniques are shown to be dependent on cultural narratives prevalent in their wider social setting. On the historical level, the project is an investigation of the cultural roots of infinitesimal methods in early modern England. The central argument, on all levels, is that certain mathematical techniques, developed by Elizabethan mathematical practitioners, were shaped by a ubiquitous cultural narrative.

A central focus of this book is the work of Thomas Hariot, and it serves to demonstrate the close interaction of narrative and mathematical practices. Hariot was probably the most original English mathematician before the Interregnum, as well as being an explorer in his own right. The discussion in Chapters 4 and 5 shows how a simple tale of exploration and discovery, prevalent in Elizabethan mathematical circles, helped structure his fundamental approach to the classical problems of infinitesimals and the mathematical continuum. Hariot was known to his contemporaries for his atomism, which extended from his views on matter to the structure of the mathematical continuum. His main contention was that seemingly smooth and homogeneous magnitudes were in fact composed of a finite or infinite number of discrete parts. A line, for instance, is construed as a conglomeration of points, a surface is composed of lines, and so on. Hariot developed this approach despite his awareness that it violated the basic tenets of classical geometry and led to paradoxes and inconsistencies. The paradoxes, Hariot claimed, merely obscure the true hidden nature of the geometrical continuum. A mathematician, he argued, must break through the confusing paradoxes and uncover the true nature of the continuum. Hariot may well have been right: the mathematical approach he advocated was pivotal in the development of logarithms and the early forms of the calculus.

Significantly, the exploration literature of Hariot's time was promoting a vision of English expansion and conquest that was remarkably similar to his mathematical approach. The imperialist rhetoric was characterized by a standard narrative, which was applied with regularity to the different voyages and ventures. Time after time the voyagers and their chroniclers repeated the tale of a magical kingdom deep within the land, rich in gold and treasure. The land, they claimed, is well hidden and naturally protected and appears to be completely inaccessible to the traveler who arrives on its hostile coast. The enterprising explorer, however, must not despair: on closer inspection he will discover that rivers and other passages break up the smooth seashore and formidable mountain ranges and provide passage to the interior. The traveler will enter the maze of passageways, overcome all obstacles, and find his way through to the golden land.[21]

The Elizabethan mathematical practitioners were no strangers to the imperialist project. Hariot and his colleagues designed nautical instruments, calculated the astronomical tables necessary for oceanic navigation, and some, including Hariot, joined the expeditions themselves. Even more importantly, the mathematicians took an active part in fashioning and utilizing the rhetoric of the enterprise: they composed promotional pamphlets for the voyages and drew maps of the geographical discoveries, which accorded with the standard narrative of exploration. The golden land, the natural obstacles, and the breaks in between were all present in the explorers' maps drawn by Elizabethan mathematicians.

It is hardly surprising, then, that we find the mathematical practitioners applying the same rhetoric to their own professional interests. Hariot and his colleagues start referring to mathematics itself as unexplored territory, possessing great treasures, but unapproachable to the uninitiated. The golden land of knowledge is obscured in these accounts by various technical difficulties, which correspond to the natural obstacles of the geographical tale. The mathematicians then proceed to fashion themselves as enterprising explorers, showing the way to the hidden secrets of their field.

Hariot went further than others. Although some of his colleagues used the rhetoric of exploration as a guiding metaphor for their pursuit of knowledge, Hariot utilized the discourse in defining his technical mathematical approach. The paradoxical mathematical continuum, for Hariot, was no different than the impenetrable coasts of the undiscovered lands. The manner to overcome the difficulties was, therefore, the same in both cases: the explorer must locate the elusive passages which break down the continuous (coast)line and find the way to the hidden secrets within. In the process, the smooth impenetrable continuum was transformed into a conglomeration of discreet parts, separated by regularly placed cleavages. The continuum became what Hariot called a "discretum," thus opening the way for the indivisiblist techniques that preceded the calculus in the seventeenth century.

Hariot treated the problem of the continuum in the same manner as he approached the geography of undiscovered lands. In both cases he posited great wealth hidden beyond confusing and seemingly impenetrable obstacles. In both cases he then proceeded to locate the breaks in the protective bulwark and thus provide access to the riches within. In his map of Guyana Hariot sketched regularly placed rivers, which penetrated the mountains into the golden plain of El Dorado. Similarly, when encountering the seemingly continuous mathematical landscape with its attendant paradoxes and confusions, Hariot chose the same approach: the "mystery of infinites," as he called it, was solved by dividing the continuum into discrete parts by regularly placed breaks. In mathematics, as also in cartography, Hariot's work was guided by the standard narrative of exploration and discovery.

Just like Tolkein or Frank Herbert in Restivo's account, Hariot had created a self-contained mathematical world. It was ruled by the laws of mathematical reasoning, obscured by the confusions and contradictions of ancient paradoxes, and clarified, eventually, through its division into parts. In the end, Hariot had produced a unique mathematical landscape in which the smooth continuum was broken by regularly placed openings. It is no coincidence that Hariot's mathematical space unmistakably resembled his geographical charts of undiscovered lands and his vision of the inner structure of matter. All three were structured by the same story—the simple tale of exploration and discovery that pervaded imperialist circles in Elizabethan England.

Appendix B

HARIOT'S METHOD OF CALCULATING MERIDIONAL PARTS

Hariot approached his analysis of rhumb lines on a sphere methodically.[1] He first set out to prove that the stereographic projection from the surface of a sphere to its horizontal plane is conformal. This means that if an angle between two lines is given on a sphere, and the lines are then projected onto its equatorial plane, the angle between the projected lines on the plane would be equal to the original spherical angle. This preliminary stage was crucial for Hariot's project, since it allowed him to project the spherical rhumb onto the flat surface of a map. The spherical rhumb is defined as a path that always maintains the same bearing. In effect, it crosses all meridians at a fixed angle. Since the stereographic projection preserves these angles, the rhumb will be projected onto the equatorial plane as a spiral that intersects with all lines radiating from the central pole (the radius vectors) at the same fixed angle.

Another characteristic of the stereographic projection is that all latitude lines from the sphere are projected on the equatorial plane as concentric circles around the central pole. Through simple geometry Hariot found that if the radius of the sphere is r_0, then the radius of the circle projected onto the equatorial plane from latitude ϕ is

$$\tan\left(\frac{\pi}{4} - \frac{\phi}{2}\right) r_0.$$

Since the rhumb progressively crosses all latitude circles on the sphere, Hariot concluded that the radius vector r of a given point on the equiangular spiral, which is the projection of a point on the sphere of latitude ϕ, is also

$$r = \tan\left(\frac{\pi}{4} - \frac{\phi}{2}\right) r_0.$$

Hariot now turned his attention to the characteristics of the equiangular spiral itself. Rather than treat it as a smooth, continuous curve, he chose to approximate it by an infinite series of similar triangles (see Figure 18 in Chapter 5). If O is the center of the equatorial plane and α the angle of the spiral (that is, the angle at which the rhumb crosses the meridians and the spiral cuts the radius vectors), and θ a fixed central angle, then

$$\Delta OAB_1 \approx \Delta OB_1 B_2 \approx \Delta OB_2 B_3 \approx \ldots \approx \Delta OB_{n-1} B_n.$$

Clearly, the smaller θ is, the better the succession of similar triangles will approximate the true spiral.

Now if we mark the radius vectors

$$OA = r_0, \, OB_1 = r_1, \, OB_2 = r_2, \, \ldots \, OB_n = r_n$$

then, because of the similar triangles we get:

$$r_1 / r_0 = r_2 / r_1 = r_3 / r_2 = \ldots = r_n / r_{n-1}.$$

In other words, the radius vectors bear a fixed ratio to each other. For any n, if we posit $r_0 = 1$, then

$$r_n = (r_1)^n.$$

This means that the values of the radius vectors of the equiangular spiral, taken at regular intervals, are simply an exponential series of r_1.

Hariot proceeded to construct an approximate spiral of $\alpha = 45°$ by calculating the radius vectors r_i at intervals of $\theta = 1'$. He did so by calculating the fixed ratio $\beta = r_i / r_{i-1}.^2$ Now, positing $r_0 = 1$, he got

$$r_1 = \beta, \, r_2 = \beta^2, \, r_3 = \beta^3 \ldots r_i = \beta^i \ldots$$

He then calculated the tables for β^i for any i. The geometrical meaning of β^i is, of course, the length of the radius vector of the equiangular spiral of 45° after an angle of i' had been turned. Since the projected radius vector of any point on the sphere of latitude ϕ was $r = \tan(\pi/4 \, \phi/2)r_0$, he now sought the proper value of "i" in the β tables for which, for a given latitude ϕ

$$\beta^{i+1} < \tan\left(\frac{\pi}{4} - \frac{\phi}{2}\right)r_0 < \beta^i.$$

For any latitude ϕ on the sphere, this "i" was the angle turned by the rhumb of 45° by the time it had reached latitude ϕ; or, to put it simply, the number of 1' meridians crossed by the 45° rhumb by the time it arrived at latitude ϕ. Since in the Mercator chart the meridians are depicted as equally spaced parallels, it would be necessary to place latitude ϕ in the location of latitude "i" if the 45° rhumb was to appear in the chart as a straight line of that bearing. In other words, "i" was the latitude to which ϕ would have to be "moved" in order to produce a chart that preserves true directions. Technically, "i" is the meridional part of latitude $\phi.^3$

Hariot's method was based on the observation that the radius vectors of the equiangular spiral were in a fixed ratio to each other, when taken at regular angular intervals. From this it follows that the length of a radius vector at a given point indicates the angle turned by the spiral up to that point. Hariot then proceeded to calculate the radius vectors of points projected from the 45° rhumb on the sphere. For any point on the circle of latitude ϕ on the sphere, this radius vector indicated the angle turned by the spiral, and hence the meridional part.

Introduction

1. On the role of the imagery of exploration in the reform of knowledge, see, for example, Anthony Grafton, *New Worlds Ancient Text: The Power of Tradition and the Shock of Discovery* (Cambridge, Mass.: The Belknap Press of Harvard University Press, 1992).

2. On the new science as a systematic "hunt" for the elusive secrets of nature, see William Eamon, "Science as a Hunt," *Physis: Rivista Internazionale di Storia della Scienza* 31 (1994): 393–432; as well as chapter 8 of William Eamon, *Science and the Secrets of Nature* (Princeton, N.J.: Princeton University Press, 1994).

3. Christopher Clavius, founder of the Jesuit mathematical tradition expresses this long-standing view in his treatise "In disciplinas mathematicas prolegomena" where he wrote, "[The mathematical disciplines] alone preserve the way and procedure of science. For they always proceed from particular foreknown principles, to the conclusions to be demonstrated."

The passage is quoted in Peter Dear, *Discipline and Experience: The Mathematical Way in the Scientific Revolution* (Chicago: University of Chicago Press, 1995), 40.

4. John Wallis in particular developed a mathematical approach he christened "the inductive method." It was based on extrapolating general mathematical laws from a few particular trials. See discussion below in Chapter 6.

5. For a discussion of the new mathematics the best source is still Carl B. Boyer, *The History of the Calculus and its Conceptual Development* (1949; reprint, New York: Dover, 1959). For a new treatment of the paradoxes and difficulties of the new mathematics, see Paolo Mancosu, *Philosophy of Mathematics and Mathematical Practice in the Seventeenth Century* (Oxford: Oxford University Press, 1996).

6. The shift from restating familiar knowledge to uncovering hidden secrets is consistent with contemporary developments in natural philosophy and the rise of empiricism. See Eamon, "Science as a Hunt."

Chapter 1

1. Columbus's account of the end of the *Santa Maria* can be found in the December 25 and 26 entries in Christopher Columbus, *The Diario of Christopher Columbus's First Voyage to America, Abstracted by Fray Bartolomé de las Casas,* ed. Oliver Dunn and James E. Kelley Jr. (Norman: University of Oklahoma Press, 1988), 279–91.

2. Columbus, *Diario,* 291, December 26, 1492.

3. Ibid.

4. For discussions of Columbus's plans for a crusade, see Tzvetan Todorov, *The Conquest of America* (New York: HarperPerennial, 1992), esp. 10–14; and Pauline Moffitt Watts, "Prophecy and Discovery: On the Spiritual Origins of Christopher Columbus's 'Enterprise of the Indies,'" *American Historical Review* 90 (1985): 73–102.

5. Christopher Columbus, "The Will of Christopher Columbus," in *The Authentic Letters of Columbus,* ed. William Eleroy Curtis (Chicago: Field Columbian Museum, 1895), 198.

6. *Raccolta di Documenti e Studi Pubblicati dalla R. Commissione Columbiana* (Rome, 1892–94), part I, vol. ii, 164–66; quoted in Todorov, *America,* 11.

7. On the *Book of Prophecies,* see Watts, "Prophecy and Discovery."

8. Christopher Columbus, *The Libro de las profecias of Christopher Columbus,* ed. and trans. Delano C. West and August Kling (Gainsville: University of Florida Press, 1991), 101.

9. Cecil Jane, ed., *The Four Voyages of Columbus* (New York: Dover, 1988), 2.

10. Stephen Greenblatt, *Marvelous Possessions* (Chicago: University of Chicago Press, 1991), 54.

11. A modern parallel of Columbus's view of exploration as a military campaign can be found in Sir Ernest Shackleton's reference to his arctic expedition as "White Warfare" in the dark days of World War I.

12. On Columbus's self-fashioning as "Christo-ferens," see, for example, Watts, "Prophecy and Discovery."

13. On the crusading ethos of the conquistadors, see, for example, Norman Housley, *The Later Crusades* (Oxford: Oxford University Press, 1992), 311–13; and J. H. Parry, *The Age of Reconnaissance* (Berkeley and Los Angeles: University of California Press, 1981), chap. 1.

14. See, for example, Bartolomé de Las Casas, *In Defense of the Indians,* trans. Stafford Poole (De Kalb: Northern Illinois University Press, 1974). The original manuscript dates from 1552.

15. Greenblatt, *Marvelous Possessions,* 53.

16. It is no coincidence that the crusades were born of the Cluny Reform of the eleventh century, which promoted a spiritual revival of the church. Significantly, Pope Urban II who launched the first crusade was a monk and prior of the Abbey of Cluny, the center of the movement. Marcus Bull points out that the Cru-

sades sought to achieve this spiritual revival through a typically medieval emphasis on "place," in this case the Holy Land. See Marcus Bull, "Origins," in *The Oxford Illustrated History of the Crusades*, ed. Jonathan Riley-Smith (Oxford: Oxford University Press, 1995); as well as the "Introduction" to Jonathan Riley-Smith, *The First Crusade and the Idea of Crusading* (Philadelphia: University of Pennsylvania Press, 1987).

17. On the various targets of crusading after the fall of the Latin East, see Norman Housley, *The Later Crusades* (Oxford: Oxford University Press, 1992).

18. On crusades that did not attempt to liberate the Holy Land, see Housley, *Later Crusades*.

19. Significantly, all crusades had to be technically "defensive," as Christians could not fight wars of conversion. See Jonathan Riley-Smith, "The Crusading Movement and Historians," in *Oxford Illustrated History of the Crusades*, 8.

20. On Columbus's search for the earthly paradise, see the "Third Voyage of Columbus" in Jane, *Four Voyages*, part 2, 2–71. Columbus's famous claim that the Earth's shape is similar to a pear, with the Earthly Paradise on its "nipple," appears on p. 30.

21. On Columbus's tendency to always "rediscover" his initial assumptions, see "Columbus as Interpreter," in Todorov, *America*.

22. For Peter Martyr's rejection of Columbus's "Earthly Paradise," see Jane, *Four Voyages*, part 2, lxxxiii; as well as Anthony Pagden, *European Encounters with the New World* (New Haven, Conn.: Yale University Press, 1992), 22. On Las Casas's comparison of America to paradise, see Mary B. Campbell, *The Witness and the Other World* (Ithaca, N.Y.: Cornell University Press, 1988), 207. On Cortés's comparison of Tenochtitlan to Jerusalem, see Jennifer R. Goodman, *Chivalry and Exploration* (Woodbridge, U.K., and Rochester, N.Y.: Boydell Press, 1998), 163–64; as well as Hernán Cortés, *Letters from Mexico*, trans. and ed. A. Pagden (New Haven, Conn.: Yale University Press, 1986), 159.

23. On romance literature and the settlement of the Americas, see Goodman, *Chivalry and Exploration*.

24. Goodman, *Chivalry and Exploration*, 154.

25. Ibid.

26. Greenblatt, "From the Dome of the Rock to the Rim of the World," in *Marvelous Possessions*, 26–51.

27. Goodman, *Chivalry and Exploration*, 49.

28. On Mandeville, see Malcolm Letts, ed., *Mandeville's Travels* (London: The Hakluyt Society Publications 2, nos. 101, 102, 1953); as well as Stephen Greenblatt, "From the Dome of the Rock to the Rim of the World," in Greenblatt, *Marvelous Possessions*.

29. For examples, see Goodman, *Chivalry and Exploration*.

30. Significantly, knights-errant may refuse to take possession of things that come their way. See Greenblatt's discussion of Mandeville's rejection of the riches of the Vale Perilous in *Marvelous Possessions*, 27.

31. See Goodman, *Chivalry and Exploration.*

32. Ibid., 151.

33. Ibid., 153.

34. Goodman makes this argument persuasively in chapter 6 of *Chivalry and Exploration*, esp. on 162.

35. The complete text of the *Historia general y natural de las Indias* remained unpublished until the nineteenth century, but a version of the first part of it appeared in Seville in 1535. For the complete text, see Gonzalo Fernandez de Oviedo Y Valdez, *Historia general y natural de las Indias*, in *Biblioteca de autores espanoles*, ed. Buesa Juan Perez di Tudela, vols. 117–21 (Madrid, 1959). Oviedo's romance of Claribalte appeared as *Libro de muy esforcado y invencible Cavallero della fortuna propriamente llamado don claribalte* (Valencia, 1519). For a discussion of Oviedo as historian and romantic author, see Pagden, *European Encounters*, 56ff.

36. Goodman, *Chivalry and Exploration*, 156.

37. A good summary of the legend of El Dorado and the various expeditions mounted to discover it in the sixteenth century can be found in the introduction to Walter Raleigh, *The Discovery of Guiana by Sir Walter Ralegh*, ed. V. T. Harlow (1595; reprint, London: Argonaut Press, 1928).

38. On the exploration of Virginia and El Dorado, see Chapter 2 below.

Chapter 2

1. Richard Hakluyt, *The Principal Navigations, Voyages, Traffiques, and Discoveries of the English Nation, in Twelve Volumes* (Glasgow: MacLehose, 1904), vol. 7, 368. See also his account of Frobisher's views on p. 364.

2. "Chart illustrating Frobisher's voyages to the north-east in 1576–78, perhaps by James Beare," in R. A. Skelton, *Explorers' Maps* (London: Spring Books 1958), fig. 75.

3. *The Principal Navigations*, 7:391.

4. Ibid., 7:392.

5. Ibid., 7:400.

6. Ibid., 7:405.

7. Ibid., 7:443.

8. The horn of a "sea unicorn" found by Frobisher's men, for instance, was brought back and displayed at Windsor. See Samuel Purchas, *Purchas his Pilgrimes*, vol. 14 (London, 1625), 399; quoted in Robert Ralston Cawley, *Unpathed Waters* (Princeton, N.J.: Princeton University Press, 1940), 70. Frobisher also captured several natives and brought them to England for display.

9. On the views of Amadas and Barlowe, Thomas Hariot, and Henry Briggs on the geography of Virginia, see discussion below.

10. See Michael Lok's world map, published by Hakluyt in 1582, in Skelton, *Explorers' Maps*, fig. 76.

11. It is, of course, extremely doubtful whether any account can be considered "simply a reflection of experience." My purpose here is not to contrast the skewed mediated perspective of the imperialists with the true view of an "innocent eye," but rather to trace how one particular brand of mediation shapes its own worldview.

12. Stephen Greenblatt discusses at length how explorers' foreknowledge of what they would find determined their discoveries. Columbus repeatedly observes signs that the great civilizations of the East are close by and confidently interprets the natives as telling him so. Frobisher is similarly confident of his knowledge of Baffin Island and his understanding of the natives. In both cases this misplaced confidence clashes with the unyielding realities of the unknown lands. See "Kidnapping Language," in Stephen Greenblatt, *Marvelous Possessions* (Chicago: University of Chicago Press, 1991). Tzvetan Todorov makes a similar point in his discussion of Columbus's interpretations of the new lands. See Tzvetan Todorov, *The Conquest of America* (New York: HarperPerennial, 1984), chap. 2. My own discussion attempts to show how the coherent and tightly formulated set of expectations of English explorers resulted in a particular topographical account when it came into contact with the realities of alien shores.

13. For Barthes's discussion of the nature of myths as well as the specific example of the magazine cover, see Roland Barthes, "Myth Today," in Barthes, *Mythologies*, ed. and trans. Annette Lavers (New York: Noonday Press, 1972), original French published in Paris, 1957.

14. Dennis Wood argues forcefully that maps serve interests by naturalizing power relationships. He identifies maps as Barthian "myths," which transform an ideological position into a self-evident, natural "truth." See Dennis Wood, *The Power of Maps* (New York: Guilford Press, 1992), esp. chap. 5; and Barthes, "Myth Today." For a related argument about natural landscape painting as the "universal language" of imperialism, see W. J. T. Mitchell, "Imperial Landscape," in *Landscape and Power*, ed. W. J. T. Mitchell (Chicago: University of Chicago Press, 1994). Much has been written in recent years about the ways in which maps both embody cultural norms and facilitate the establishment of power relationships. David Turnbull's *Maps Are Territories* (Chicago: University of Chicago Press, 1993) summarizes these arguments in concise form. J. B. Harley argues that maps are ideological tools and symbols; see his "Silences and Secrecy: The Hidden Agenda of Cartography in Early Modern Europe," *Imago Mundi* 40 (1989): 58–76. For an earlier form of Harley's argument, see his "Meaning and Ambiguity in Tudor Cartography," in *English Map Making 1500–1650*, ed. Sarah Tyacke (London: The British Library, 1983). For a general argument about the role of maps as shapers of worldviews and as guides to action, see Wilbur Zelinsky, "The First and Last Frontier of Communication: The Map as Mystery," in the Special Libraries Association, *Geography and Map Division Bulletin* no. 94 (December 1973): 2–8. Richard Helgerson makes a fascinating argument about the changing political role of maps in Elizabethan and early Stuart England in his "The Land Speaks: Cartography, Chorography, and Subversion in Renaissance

England," *Representations* 16 (fall 1986). A version of the article later appeared as a chapter in Richard Helgerson, *Forms of Nationhood: The Elizabethan Writing of England* (Chicago: University of Chicago Press, 1992). According to Helgerson, while Elizabethan mapping represented royal power over the land, under the Stuarts it came to express the land "in itself," independent of royal power. Helgerson emphasizes the nontopographical aspects of the maps as carrying a political message (such as the gradual marginalization of the royal arms on the maps), while my argument emphasizes the actual contours of the land depicted. A general account of mapping practices in early modern Europe, arranged by country, is David Buisseret, ed., *Monarchs, Ministers, and Maps: The Emergence of Cartography as a Tool of Government in Early Modern Europe* (Chicago: University of Chicago Press, 1992).

15. For a concise account of the Frobisher and Davis voyages, see K. R. Andrews, *Trade, Plunder and Settlement* (Cambridge: Cambridge University Press, 1984), chap. 8. The maps can be found in P. Cumming, R. A. Skelton, and D. B. Quinn, *The Discovery of North America* (New York: American Heritage Press, 1972).

16. Even this small complement did not last long, as the *Michael* returned to England on its own accord, and the pinnace sank before ever reaching the American coast. For a brief account of the voyages, see Andrews, *Trade, Plunder*, chap. 8. For more detailed accounts and documentation, see Richard Collinson, *Three Voyages of Martin Frobisher* (London: The Hakluyt Society Publications 38, 1867); and Vilhjalmur Stefansson and Eloise McCasskill, eds., *The Three Voyages of Martin Frobisher*, 2 vols. (London: Argonaut Press, 1938).

17. *The Principal Navigations*, 7:362, in Best's account.

18. For more detailed accounts of Davis's voyages, see Andrews, *Trade, Plunder*; Albert H. Markham, *The Voyages and Works of John Davis the Navigator* (London: The Hakluyt Society Publications 59, 1880); John Janes's account of the voyages in Hakluyt's *Principal Navigations*; and Davis's own remarks in *The Worldes Hydrographical Description* (London, 1595).

19. "A Letter of Sir Humphrey Gilbert Knight, Sent to his Brother, Sir John Gilbert, of Compton, the last of June, 1566." The letter is printed in the preface to Gilbert's *A Discourse of the Discoverie for a New Passage to Cathaia* (London, 1576). The tract itself was, however, written in 1566. See David B. Quinn, *The Voyages and Colonising Enterprises of Sir Humphrey Gilbert* (London: The Hakluyt Society Publications 2, nos. 83–84, 1940), 134–35. All page references are from this edition.

20. Indeed, Mandeville's *Travels* was one of only nine books in the library of Frobisher's expedition. See Andrews, *Trade, Plunder*, 171.

21. Andrews, *Trade, Plunder*, 171.

22. B.L. Cotton MS Otho. E. VIII, f. 46. Published in Collinson, *Three Voyages*, 79.

23. Collinson, *Three Voyages*, 19.

24. Both Lok and Frobisher suggest on occasion that the expedition did in fact arrive in Asia. According to Best, when Frobisher entered Frobisher's strait "that land upon his right hand as he sailed Westward he judged to be the continent of Asia, and there to be divided from the firme of America, which lieth upon the left hand over against the same." See *The Principal Navigations*, 8:280. For Lok's views, see Andrews, *Trade, Plunder*, 172. But whether the gold of Asia is transferred to Baffin Island or the island itself *becomes* Asia, the result is the same: Meta Incognita supplants Cathay as the target of the expedition and the source of great riches.

25. *The Principal Navigations*, 7:283.

26. B.L. Cotton MS Otho. E. VIII, f. 45, printed in Markham, *Frobisher*, 91.

27. Collinson, *Three Voyages*, 91–102.

28. *The Principal Navigations*, 7:217. This and all other references to Settle can also be found in Dionyse Settle, *Laste Voyage into the West and Northwest Regions* (London, 1577).

29. *The Principal Navigations*, 7:228.

30. Ibid., 7:229.

31. Ibid., 7:289.

32. Ibid., 7:296.

33. Ibid., 7:214.

34. Ibid., 7:232.

35. Ibid., 7:218.

36. Ibid., 7:233.

37. Ibid., 7:279.

38. Ibid., 7:288.

39. Ibid., 7:289.

40. Ibid., 7:282.

41. Ibid., 7:291.

42. Ibid., 7:340.

43. Ibid., 7:402.

44. Ibid., 7:335.

45. Ibid., 7:328.

46. A 1974 expedition identified Frobisher's "black ore" as calc-silicate gneiss, and the "red ore" as metasandstone. See the appendix to W. A. Kenyon, *Tokens of Possession: The Northern Voyages of Martin Frobisher* (Toronto: Royal Ontario Museum, 1975). The "ore" was eventually used to repair the roads in the vicinity of Bristol.

47. For Settle's descriptions of Friezland, see *The Principal Navigations*, 7:214, 232.

48. On Columbus's practice of reading signs in accordance with his preconceived notion, see Todorov, *America*, esp. the section "Columbus as Interpreter." See also Greenblatt, *Marvelous Possessions*, chap. 4.

49. *The Principal Navigations*, 7:214–15.

50. Ibid., 7:279–80.

51. Best's characterization of the Strait of Magellan does not accord well with the actual geography of the region, but it does agree perfectly with the description offered by Richard Hakluyt the Younger in his proposal to colonize the straits in 1579. See E. G. R. Taylor, ed., *The Original Writings and Correspondence of the Two Richard Hakluyts*, vol. 1 (London: The Hakluyt Society Publications 2, no. 76, 1935), doc. 24, 139–46. Taylor's collection is cited below as "*Two Hakluyts.*"

52. *The Principal Navigations*, 7:336.

53. "The first voyage of M. John Davis, undertaken in June 1585, for the discoverie of the North-west passage, Written by M. John James Marchant, sometimes servant to the worshipful Master William Sanderson," in *The Principal Navigations*, 7:390.

54. John Davis, *Worldes Hydrographical Description*. Printed in Albert Hastings Markham, ed., *Voyages and Works of John Davis the Navigator* (London: The Hakluyt Society Publications 59, 1880), 191–228. The quote is on p. 210.

55. *The Principal Navigations*, 7:215.

56. Ibid., 7:233.

57. This passage is particularly significant since what seems as a liberating breakthrough in the quote, turns into a threatening situation by the end of the sentence, and into a death trap later in the paragraph: "The yce being round about us and inclosing us, as it were within the pales of a parke. . . . But the storme so increased, and the waves began to mount aloft which brought the yce so neare us, that we were faine to beare in and out. . . . Thus the yce comming on us so fast, we were in great danger, looking every houre for death." See *The Principal Navigations*, 7:234. Note that the familiar trope that posits a liberating passage to the interior is casually used here despite the life-threatening reality of the situation.

58. *The Principal Navigations*, 7:236.

59. Ibid., 7:237.

60. Ibid., 7:340.

61. Ibid., 7:338.

62. Ibid., 7:339.

63. Ibid.

64. Abraham Fleming, "A Rythme Decasyllabicall, upon this last luckie voyage of worthie Capteine Frobisher, 1577," in Settle, *Last Voyage into the West.*

65. See, for instance, Andrews, *Trade, Plunder*, 177.

66. See, for instance, Richard Hakluyt the elder's "Pamphlet for the Virginia Enterprise, 1585," in E. G. R. Taylor, ed., *Two Hakluyts*, vol. 2 (77), doc. 47, p. 328, where Hakluyt suggests that the voyage to Virginia will lead to a discovery of passages to "Iapan, China, and Cathay."

67. The fullest documentation can be found in David B. Quinn, *The Roanoke Voyages 1584–1590* (The Hakluyt Society, 1955; reprint, Nendeln, Liechtenstein: Kraus Reprint, 1967). The Nendeln/Liechtenstein edition is cited below as *The*

Roanoke Voyages. A detailed account based on the documents is David B. Quinn, *Set Fair for Roanoke* (Chapel Hill: University of North Carolina Press, 1984). Andrews's *Trade, Plunder and Settlement* contains a brief account of the expeditions.

68. *The Principal Navigations*, 7:290.

69. Thomas Hariot, *A Briefe and True Report of the New Found Land of Virginia* (New York: Dover Publications, 1972). This edition is a reproduction of the first volume of Theodor de Bry's *America* series. See Theodor de Bry, ed., *America* (Frankfurt am Main, 1590). The text was originally published in London as a short pamphlet in 1588. For a complete account of the American work of John White, including reproductions of all known watercolors and engravings, see Paul Hulton and David B. Quinn, *The American Drawings of John White*, 2 vols. (London: The British Museum, 1964). All the documentary evidence on the Virginia colonies was gathered by David B. Quinn and published in *The Roanoke Voyages*.

70. *The Roanoke Voyages*, doc. 4, 95.

71. Ibid., doc. 4, 98.

72. Ibid., doc. 4, 106. Wayne Franklin proposes three types of early American travel narratives: "discovery narratives," full of wonder at the new land; "exploration narratives," which seek to incorporate the land's abundance into a familiar framework; and "settlement narratives," born out of the actual encounter with the harsh realities of life in the land. Amadas and Barlowe's report, imbued with wonder and awe at the new land, is typical of Franklin's exploration phase. See Wayne Franklin, *Discoverers, Explorers, Settlers* (Chicago: University of Chicago Press, 1979).

73. *The Roanoke Voyages*, doc. 24, 199.

74. Ibid., doc. 28, 208.

75. On Hariot's *Briefe and True Report*, see n. 69 above.

76. Hariot, *Briefe and True Report*, 31.

77. Ibid.

78. See, for instance, the list of commodities to be looked for in Richard Hakluyt the elder, "Inducements to the Liking of the Voyage Intended towards Virginia, 1585," in E. G. R. Taylor, ed., *Two Hakluyts*, doc. 47; Richard Hakluyt the younger, "Epistle Dedicatory to Sir Walter Raleigh, 1587," in *Two Hakluyts*, doc. 58 (esp. p. 378); Richard Hakluyt the younger's letter to Raleigh of December 30, 1586, printed in *The Roanoke Voyages*, doc. 69, 494, in which he mentions silver mines in the interior; and Lane's description of the land of Chaunis Temoatan in *Two Hakluyts*, doc. 45, 259. The image predated Raleigh's expeditions, as is evidenced by Richard Hakluyt the younger, "Discourse of Western Planting, 1584," printed in *Two Hakluyts*, doc. 46, and persisted when the enterprise was revived later on, as can be seen in Richard Hakluyt the younger, "Epistle Dedicatory to Sir Robert Cecil, 1599," in *Two Hakluyts*, doc. 76, 456; and Richard Hakluyt, "Epistle Dedicatory to the Council of Virginia, 1609," in *Two Hakluyts*, doc. 89, 500–1.

79. Hariot, *Briefe and True Report*, 10.

80. *Two Hakluyts*, doc. 47, 330.

81. The gendered aspects of the positioning of the land as possessing hidden secrets and awaiting her discoverer are discussed in Patricia Parker, "Rhetorics of Property: Exploration, Inventory, Blazon," in her *Literary Fat Ladies* (London: Methuen, 1987); and Annette Kolodny, *The Lay of the Land* (Chapel Hill: University of North Carolina Press, 1975).

82. *The Roanoke Voyages*, doc. 4, 93–94.

83. See n. 69 above. On Hariot's mapping activities, see Amir R. Alexander, "Lunar Maps and Coastal Outlines: Thomas Hariot's Mapping of the Moon," *Studies in the History and Philosophy of Science* 29 (1998): 345–68.

84. Hariot, *Briefe and True Report*, 45.

85. *The Roanoke Voyages*, 414, n. 5.

86. The documents relating to the English colonization of Virginia are full of references to the enterprise in terms of the master narrative of exploration. Examples abound: Richard Hakluyt the younger suggests in his "Discourse on Western Planting" of 1584 that "Ryvers all along many hundreth miles into the Inland are infinitely full fraught" with various valuable commodities (*Two Hakluyts*, doc. 46, 280–82; the quote is from p. 281). Richard Hakluyt the elder makes the same point in his 1585 tract "Inducements to the Liking of the Voyage Intended towards Virginia" (*Two Hakluyts*, doc. 47, esp. article 16, pp. 329 and 334). In his 1584 pamphlet of the same name he commended Virginia for her deep rivers that enable effective control of the interior and suggested the possibility of a northwest passage through Virginia (*Two Hakluyts*, doc. 48, 341–42; for quote see n. 120). Ralph Lane, in a letter to Walsingham from Virginia, referred to the land as both incredibly plentiful and "by Nature fortefyed to ye sea ward" (*The Roanoke Voyages*, doc. 24, 200). The passages to the interior also play a role in Lane's letter, as when he quotes Hariot that the river of Moratico may prove to be a convenient passage to the South Sea (*The Roanoke Voyages*, doc. 45, 273–74; see n. 120). The persistence of the tradition can be detected in Hakluyt the younger's Epistle Dedicatory to the Council of Virginia in 1609, where he repeated the rumors of the existence of gold in the Virginian interior and assured the prospective investors that Hariot had reliable information about it from the Indians (*Two Hakluyts*, doc. 89, 500–501). The importance of the passages leading to the golden interior is stressed in his 1606 instruction to the Virginia colony (*Two Hakluyts*, doc. 87, 493–94).

87. See discussion above as well as Barthes's "Myth Today." As Barthes points out, the first-level sign, in this case the map of the Carolina Banks, serves to naturalize the larger ideological claim, which is here the master narrative of exploration.

88. *The Roanoke Voyages*, doc. 45, 259–60.

89. Ibid., doc. 45, 270–72.

90. A good summary of the legend of El Dorado and the various expeditions mounted to discover it can be found in the introduction to Walter Raleigh, *The Discovery of Guiana by Sir Walter Raleigh*, ed. V. T. Harlow (London: Argonaut Press,

1928). For a participant account of Ursua and Aguirre's expedition, see William Bollaert, *The Expedition of Pedro de Ursua & Lope de Aguirre in Search of El Dorado
and Omagua in 1560–1* (London: The Hakluyt Society Publications 28, 1861).

91. For a detailed account of Raleigh's contacts with Gamboa, his interest in
El Dorado, and his 1595 expedition, see Charles Nicholl, *The Creature in the
Map* (New York: William Morrow & Company, 1996). On Raleigh's meeting
with Gamboa, see also Andrews, *Trade, Plunder*, 287.

92. Raleigh, *Discovery*.

93. Ibid., 54.

94. Ibid., 51–52.

95. Keymis's full account of his voyage is found in Lawrence Keymis, *A Relation of the Second Voyage to Guiana* (London, 1596; reprint, Amsterdam: Theatrum Orbis Terrarum, 1968). A brief account of Raleigh's and Keymis's expeditions can be found in Andrews, *Trade, Plunder*. For a literary analysis of the texts
associated with Raleigh's and Keymis's expeditions to El Dorado, see Mary C.
Fuller, "Raleigh's Fugitive Gold: Reference and Deferral in *The Discoverie of
Guiana*," *Representations* 33 (1991).

96. The actual exploration did, of course, have to contend with the geographical features of Guyana. This, however, would have little effect on the perceived
features of El Dorado. If the geography of the land would radically diverge from
the one predicted for Manoa, the inevitable conclusion would be that El Dorado
is someplace else rather than in Guyana. The geography posited for El Dorado
would, in any case, remain unchanged.

97. Raleigh, *Discovery*, 71.

98. Keymis, *Second Voyage*, C3r. "Raleana" was the name given by the English to the Orinoco.

99. Keymis, *Second Voyage*, E2v.

100. Raleigh, *Discovery*, 73. The reference to Guyana as having her "mayden-
head" is reminiscent of the land that engaged Raleigh's interests earlier—namely,
Virginia—defined as virgin land by its very name. The arrangement of two sym-
metrical forts keeping watch over the passage brings to mind the two islands
guarding the entrance to Frobisher's strait in Settle's account. See *The Principal
Navigations*, 7:214–15.

101. Raleigh, *Discovery*, 73.

102. Keymis, *Second Voyage*, F2v.

103. On the theme of a land, gendered as female, and therefore both closed
and open, see Peter Stallybrass, "Patriarchal Territories: The Body Enclosed," in
Rewriting the Renaissance, ed. Margaret W. Ferguson, Maureen Quilligan, and
Nancy J. Vickers (Chicago: University of Chicago Press, 1986), chap. 7.

104. For discussions of gendered narratives of exploration, see Parker,
"Rhetorics of Property"; Kolodny, *Lay of the Land*; and Mary Louise Pratt,
Imperial Eyes (London: Routledge, 1992). For an analysis of the textual tension in
Raleigh's *Discovery*, see Fuller, "Raleigh's Fugitive Gold."

105. Keymis, *Second Voyage*, B4r.

106. Ibid., B2r.

107. Ibid., B4r+v.

108. Ibid., E3r.

109. There are many other places in the literature associated with the Guyana voyages that point to the structuring presence of the master narrative of exploration. In his dedication of the *Discovery*, for example, Raleigh significantly insists that the true riches of the Spanish Empire lie not on the coast but in the interior (Raleigh, *Discovery*, 5–6). In praising Guyana in the conclusion, he states that "there is no countrey which yeeldeth more pleasure to the Inhabitants . . . as *Guiana* doth," and he follows this by listing the natural riches of the land (Raleigh, *Discovery*, 71–72). Raleigh uses similarly superlative language in describing Guyana to Sir Robert Cecil in his letter dated November 13, 1595. In another letter to Cecil (now Lord Salisbury), in 1607, Raleigh vouches for the existence of a rich mine in Guyana, located in a mountain and close to a river. He repeats the same description in a 1612 letter to the lords of the council. See Edward Edwards, *The Life of Sir Walter Raleigh Together with his Letters* (London: Macmillan & Co., 1868), vol. 2, letters LV (p. 109), CLXV (p. 389), and CXLVIII (p. 338). Theodor de Bry reproduced Raleigh's *Discovery* in part 8 of his *America* series, published in 1599. He added several plates, including a description of how the Indians gather golden grains from the lake and rivers of El Dorado and mold them into images. George Chapman's poem "De Guiana Carmen," printed in the preface to Keymis's *Second Voyage*, follows the general outline of the master narrative and promotes an adventurous spirit for the enterprise. See discussion below in Chapter 4.

110. The map is catalogued as B.L. Add. MS 17940. It may well be the "large Chart or Map" that Raleigh mentions he is preparing in the *Discoverie* (Raleigh, *Discovery*, 25.) The attribution to Hariot is based on a letter by Hariot to Sir Robert Cecil preserved in Hatfield House (Hatfield MS CP 42/36) dated July 11, 1596, in which Hariot proposes to add Keymis's discoveries to a map of Guiana he has prepared based on Raleigh's report. The letter is printed in John W. Shirley, *Thomas Harriot: A Biography* (Oxford: Clarendon Press, 1983), 230–32. Walter Raleigh also mentions a "plott" of Guyana prepared by Hariot in a letter to Sir Robert Cecil dated November 13, 1595. See Edwards, *Life of Sir Walter Raleigh*, vol. 2, letter LV, pp. 109–11. The attribution of the map to Hariot is, of course, a natural one, as he was the most qualified practicing mathematical practitioner in Raleigh's circle. Hariot's manuscripts in the British Library contain a sketch of a map of the island of Trinidad and the Guyana coast, in B.L. Add. MS 6786, f. 464. For a discussion of Hariot's maps as well as the coastal sketches that can be found among his papers, see Alexander, "Lunar Maps and Coastal Outlines."

111. A similar map, showing the mountains, the lake, and the rivers leading to it, can be found in de Bry's account of Raleigh's search for El Dorado in the 1634 German edition of his *America*, part 8. The map is reproduced in Theodor de Bry,

Discovering the New World, Based on the Work of Theodore de Bry, ed. Michael Alexander (New York: Harper & Row, 1976), 172.

112. The original is in Keymis, *Second Voyage*, A4v. The above translation follows closely, although not in all detail, the one given by Muriel Rukeyser, *The Traces of Thomas Hariot* (New York: Random House, 1970), 140.

113. The Guyana project was not abandoned, however, and was carried on by private entrepreneurs on a reduced scale. Robert Harcourt traveled there in 1613, as did Walter Raleigh famously in 1617 in the voyage that led to his execution. See Robert Harcourt, *A Relation of a Voyage to Guiana* (London: The Hakluyt Society Publications 2, 1928); and Vincent T. Harlow, *Raleigh's Last Voyage* (London: Argonaut Press, 1932).

114. A comprehensive list of the various expeditions can be found in Cliff Holland, *Arctic Exploration and Development: An Encyclopedia* (New York: Garland Publishing, 1994).

115. Henry Briggs, "A Treatise of the Northwest Passage to the South Sea, through the Continent of Virginia and by Fretum Hudson," appended to Edward Waterhouse, *A Declaration of the State of the Colony and Affaires in Virginia* (London, 1622), 45–50. Briggs's treatise, along with his map of North America, was later republished in Samuel Purchas, *Purchas his Pilgrimes*, book 4 (London, 1625), 852–53.

116. For a detailed account of mathematicians' careers during this period, see Mordechai Feingold, *The Mathematicians' Apprenticeship: Science, Universities and Society in England, 1560–1640* (Cambridge: Cambridge University Press, 1984).

117. William Gilbert, *De Magnete* (London, 1600), book 5. Gilbert's views were later shown to be erroneous. The magnetic dip was never used in actual navigation.

118. See Thomas Blundeville, *The Theoriques of the Seven Planets* (London, 1602); and Marke Ridley, *A Short Treatise of Magneticall Bodies and Motions* (London, 1613).

119. Wright published the tables in his *Certaine Errors in Navigation* (London, 1599). Mercator projection maps preserve "true directions"—that is, the path of a ship steering at a fixed bearing appears on a Mercator map as a straight line at the given angle. The Mercator problem and Wright's solution to it are discussed extensively below in Chapter 5.

120. Napier's book was first published in Latin as John Napier, *Mirifici Logarithmorum Canonis Descriptio* (Edinburgh, 1614). Wright's translation was published posthumously by Briggs as John Napier, *A Description of the Admirable Table of Logarithms* (London, 1616). For a discussion of Wright's translation of Napier, see Chapter 3 below.

121. Our knowledge of Briggs's involvement in the 1631 voyages is due entirely to Luke Foxe's account in his 1635 report of his voyage entitled *North-West Fox*. The accounts of both Foxe and James have been published by the Hakluyt society as Miller Christy, ed., *The Voyages of Captain Luke Foxe of Hull and*

Captain Thomas James of Bristol (London: The Hakluyt Society Publications, nos. 88–89, 1894), see esp. 262–63.

122. Thomas James's "Briggs His Bay" is listed in Christy, *Voyages of Foxe and James*, clxxvi. Luke Foxe recounts his naming of "Briggs His Mathematickes" in section 30 of *North-West Fox*. See Christy, *Voyages of Foxe and James*, 2:329.

123. See n. 115 above.

124. Thomas Hariot had apparently preceded Briggs in suggesting the possibility of a river passage to the South Sea through Virginia. In a 1585 letter to Sir Francis Walsingham, Ralph Lane wrote that the "river of Moratico promiseth great things, and by the opinion of Master Harriots the heade of it by the description of the countrey, either riseth from the bay of Mexico, or else very neare unto the same that openeth out into the South sea" (*The Roanoke Voyages*, doc. 45, 273–74). Richard Hakluyt the elder made a less specific claim as early as 1584 in his pamphlet "Inducements to the Liking of the Voyage Intended towards Virginia" where he suggested that "there is great hope to sayle into the south sea" from the north parts of Virginia (*Two Hakluyts*, doc. 48, 341–42).

125. Briggs, "Northwest Passage," 45–46.

126. This is, undoubtedly, a geographically questionable point. In fact, the West Indies served as an intermediary point for many Virginia voyages, rather than the other way around.

127. Briggs, "Northwest Passage," 46.

128. Ibid.

129. Ibid., 47. The "Fals" are most likely Niagara Falls, although there is no textual evidence to fix that with certainty.

130. Briggs, "Northwest Passage," 47.

131. Ibid., 48.

132. Ibid.

133. Ibid., 50.

Chapter 3

1. B.L. Add. MS 6788, f. 490. The poem is also quoted with variations, due to the difficult handwriting, in Muriel Rukeyser, *Traces of Thomas Hariot* (New York: Random House, 1970), 256–57; and in Jon V. Pepper, "Hariot's Earlier Work on Mathematical Navigation: Theory and Practice," in *Thomas Harriot, Renaissance Scientist*, ed. John W. Shirley (Oxford: Clarendon Press, 1974), 59.

2. Robert Recorde, *The Whetstone of Witte* (London, 1557).

3. For an account of Dee's activities, see E. G. R. Taylor, *Tudor Geography* (London: Methuen, 1930).

4. For Wright's work on logarithms, see John Napier, *A Description of the Admirable Table of Logarithms*, trans. Edward Wright (London, 1616). A modern facsimile edition was published by De Capo Press, Theatrum Orbis Terrarum

in Amsterdam, 1969. On the Mercator projection, see Edward Wright, *Certaine Errors in Navigation, Arising either of the Ordinarie Erroneous Making or Using of the Sea Chart, Compasse, Crosse Staffe, and Tables of Declination of the Sunne, and Fixed Starres Detected and Corrected* (London, 1599). On meridional parts, see detailed discussion below in Chapter 5.

5. For Hariot's connection to Elizabethan imperialism, see David B. Quinn, "Thomas Harriot and the New World," in Shirley, *Thomas Harriot, Renaissance Scientist*. For Hariot's work on mapmaking and navigation, see Jon V. Pepper, "Harriot's Calculation of the Meridional Parts as Logarithmic Tangents," *Archive for the History of Exact Sciences* 4 (1967–68): 359–413; and Pepper, "Harriot's Earlier Work." Hariot's notes on shipbuilding and ship handling can be found in B.L. Add. MS 6788, ff. 1–48, passim. Part of Hariot's work will be discussed in far greater detail in Chapter 5 below.

6. E. G. R. Taylor, *The Mathematical Practitioners of Tudor and Stuart England* (Cambridge: Cambridge University Press, 1954); E. G. R. Taylor, *Tudor Geography 1485–1583* (London: Methuen, 1930); and E. G. R. Taylor, *Late Tudor and Early Stuart Geography 1583–1650* (London: Methuen, 1934).

7. For such tracts, see the "Works" section in Taylor, *Mathematical Practitioners*; and D. W. Waters, *The Art of Navigation in England in Elizabethan and Early Stuart Times* (New Haven, Conn.: Yale University Press, 1958).

8. Hariot's manuscripts in the British Library include many folios on topics such as shipbuilding, watch-keeping on an oceangoing ship, the proper length of masts, nautical knots, and other technical issues. See B.L. Add. MS 6788, ff. 1–48. Hariot also collected maps of America and navigational rutters for Sir Walter Raleigh. On Hariot's map collecting, see David B. Quinn, ed., *The Hakluyt Handbook* (London: The Hakluyt Society Publications 2, no. 144, 1974), 50. One of Hariot's rutters survives as B.L. Sloane MS 2292. For discussions of Hariot's work as a technical consultant to the voyages, see E. G. R. Taylor, "Hariot's Instructions for Raleigh's Voyages to Guiana, 1595," *Journal of the Institute of Navigation 6* (1952): 345–51; Quinn, "Thomas Harriot in the New World," 36–53; and David B. Quinn, "Thomas Harriot and the Virginia Voyages of 1602," *William and Mary Quarterly* 3d Series, no. 27 (1970): 268–81. On Hariot's declination tables (necessary for determining latitude from the "height" of the sun at noon), see John J. Roche, "Harriot's *Regiment of the Sun* and its Background in 16th-Century Navigation," *British Journal for the History of Science* 14 (1981): 245–61. For a general account of Hariot's navigational work, see "Appendix No. 30: Thomas Hariot's Contribution to the Art of Navigation," in Waters, *Art of Navigation in England*. Jon V. Pepper's work on Hariot's calculation of tables for Mercator maps is discussed extensively in Chapter 5 below. See esp. Pepper, "Harriot's Calculation."

9. On the technical improvement involved in the great voyages, see J. H. Parry, *The Age of Reconnaissance* (London, 1963; reprint, Berkeley and Los Angeles: University of California Press, 1963), esp. part 1.

10. On the technological improvements needed for the exploration voyages, see Parry, *Age of Reconnaissance*; and Samuel Eliot Morison, *The European Discovery of America*, 2 vols. (New York: Oxford University Press, 1971). For a detailed account of navigational innovation, see Waters, *Art of Navigation in England*.

11. Dee was of Welsh descent and was therefore dissatisfied with the more obvious term "English Empire."

12. John Dee's "General and Rare Memorials" was composed of four separate tracts: the first was entitled "The Brytish Monarchie" and was published as *General and Rare Memorials Pertayning to the Perfect Arte of Navigation* (London, 1577). The second tract was a volume of navigational tables entitled "The Brytish Complement of the Perfect Art of Navigation," now lost. The third tract was burnt by Dee. The fourth, entitled "Of Famous and Rich Discoveries," partially survives in manuscript form in B.L. Cotton MS Vitellius. C. VII, ff. 25-267. A survey of Dee's tracts can be found in Nicholas H. Clulee, *John Dee's Natural Philosophy* (London and New York: Routledge, 1988). For a detailed discussion, see William H. Sherman, *John Dee: The Politics of Reading and Writing in the English Renaissance* (Amherst: University of Massachusetts Press, 1995). For an overview of Dee's imperialist activities, see Taylor, *Tudor Geography*.

13. Edward Wright, "A Cruizing Voyage to the Azores in 1589 by the Earl of Cumberland," in *A General History and Collection of Voyages and Travels*, ed. Robert Kerr, vol. 7 (Edinburgh: W. Blackwood, 1824), 375-96.

14. Henry Briggs, "A Treatise of the Northwest Passage to the South Sea, Through the Continent of Virginia and by Fretum Hudson," annexed to Edward Waterhouse, *A Declaration of the State of the Colony in Virginia* (London, 1622). Both treatises were republished in *The English Experience*, vol. 276 (Amsterdam: Theatrum Orbis Terrarum, 1970). See discussion in Chapter 2 above.

15. For Briggs's archipelago, see Edward Miller Christy, ed., *The Voyages of Foxe and James* (London: The Hakluyt Society Publications, no. 89, 1894), 329. Thomas James's "Briggs His Bay" is listed in Christy, *The Voyages of Foxe and James*, clxxvi.

16. Thomas Hariot, *A Briefe and True Report of the New Found Land of Virginia* (New York: Dover, 1972). The tract was first published in London in 1588. The Dover version is a reproduction of the much better-known edition, published by Theodor de Bry in 1590 in Frankfurt as part of his *America* series. On the circumstances surrounding Hariot's publication, see David B. Quinn, *The Roanoke Voyages 1584-1590* (The Hakluyt Society, 1955; reprint, Nendeln, Liechtenstein: Kraus Reprint, 1967).

17. A mathematical tract based on Hariot's manuscripts was published posthumously by his executors. See Thomas Hariot, *Artis Analyticae Praxis* (London, 1631). Hariot died in 1621.

18. For a detailed discussion, see Chapter 2 above. For Hariot's discussion of the coast of Guiana, see his letter to Sir Robert Cecil dated July 11, 1596, Hatfield MS CP 42/36, reproduced in Shirley, *Thomas Harriot: A Biography*, 230-31. The letter is discussed in Chapter 4 below. The map is in B.L. Add. MS 17940, and is

reproduced in Walter Raleigh, *The Discovery of Guiana by Sir Walter Raleigh*, ed. V. T. Harlow (London: Argonaut Press, 1928). The map and Briggs's tract are discussed in Chapter 2 above. For Briggs's tract, see n. 14 above.

19. John Dee, *General and Rare Memorials Pertayning to the Perfect Arte of Navigation* (London, 1577; reprint, Amsterdam and New York: De Capo Press, 1968).

20. On Dee, see Clulee, *John Dee's Natural Philosophy*; Sherman, *John Dee*; and Peter J. French, *John Dee: The World of an Elizabethan Magus* (London: Routledge, 1972).

21. John Dee, "Britanici Imperii Limites," B.L. Add. MS 59681. See discussion in Sherman, *John Dee*.

22. John Dee, "Britanici Imperii Limites," ff. 30–31.

23. Ibid., ff. 71–72.

24. Ibid., f. 21. Also discussed in Sherman, *John Dee*, 185.

25. John Dee, "Britanici Imperii Limites," f. 9.

26. Hariot, *Briefe and True Report*.

27. On Dee's frontispiece, see Lesley Cormack, "Britannia Rules the Waves? Images of Empire in Elizabethan England," *Early Modern Literary Studies* (electronic journal), September 1998, (http://www.humanities.ualberta.ca/emls/04-2 /cormbrit,html). For other discussions, see Clulee, *John Dee*, 184–85; and Frances Yates, *Astraea: The Imperial Theme in the Sixteenth Century* (London: Routledge, 1975), 49–50; French, *John Dee*, 183–85.

28. On Hariot's map, see Amir R. Alexander, "The Imperialist Space of Elizabethan Mathematics," *Studies in the History and Philosophy of Science* 26 (1995): 559–91.

29. On the impact of the discoveries on European intellectuals, see, for example, Anthony Grafton, *New Worlds Ancient Text: The Power of Tradition and the Shock of Discovery* (Cambridge, Mass.: The Belknap Press of Harvard University Press, 1992); and Anthony Pagden, *European Encounters with the New World* (New Haven, Conn.: Yale University Press, 1993).

30. Francis Bacon, *The New Organon and Related Writings* (London, 1620; reprint, New York: Liberal Arts Press, 1960), 12. William Eamon provides many other examples of early modern writers who compare natural philosophy to a voyage of geographical discovery. Sir Thomas Browne (1605–82), for example, thought that "the America and untravelled parts of truth" still await discovery, and Joseph Glanvill hoped to open up an "America of secrets and an unknown Peru of Nature. As long as we follow Aristotle, according to Glanvill, "we are not likely to reach the treasures on the other side of the *Atlantick*." See William Eamon, *Science and the Secrets of Nature* (Princeton, N.J.: Princeton University Press, 1994), 272–73.

31. Bacon, *New Organon*, 81. For a discussion of Bacon's use of geographical exploration as a model for his reform of knowledge, see Wayne Franklin, *Discoverers, Explorers, Settlers* (Chicago: University of Chicago Press, 1979), 7–10.

32. Bacon, *New Organon*, 91.

33. Ibid., 81. Bacon's papers contain a tract named "Description of the Intel-

lectual Globe," which was published only after his death. See James Spedding, Robert Leslie Ellis, and Douglas Denon Heath, eds., *The Works of Francis Bacon* (London: Longmans, 1889), 5:501–44.

34. K. R. Andrews, *Trade, Plunder and Settlement* (Cambridge: Cambridge University Press, 1984).

35. *Works of Francis Bacon*, 5:13.

36. Waters, *Art of Navigation*, chap. 1, "The Art of Pilotage."

37. See the discussion of Edward Wright's work below, as well as in Chapter 5 below.

38. For a survey of navigational techniques in this period, see Waters, *Art of Navigation*. Appendix 30 is devoted to Hariot's navigational work. For a far more detailed discussion of Hariot's work on cartography and navigation, see Pepper, "Harriot's Early Work"; and Pepper, "Harriot's Calculation."

39. Bacon, *New Organon*, 91.

40. For a recent discussion of Columbus's attitude to the New World, see Felipe Fernandez Armesto, *Columbus* (Oxford: Oxford University Press, 1991).

41. William Oughtred, *The Circles of Proportion and the Horizontall Instrument* (London, 1632). Oughtred developed the instruments many years earlier, but did not publish his invention. A short biography and a list of Oughtred's publications can be found in Taylor's *Mathematical Practitioners*; and in J. F. Scott, "Oughtred, William," in *The Dictionary of Scientific Biography*, vol. 10 (New York: Scribner, 1974). For more detailed accounts of his work, see Florian Cajori, *William Oughtred, A Great Seventeenth-Century Teacher of Mathematics* (Chicago: Open Court, 1916); Florian Cajori, "Oughtred's Ideas and Influence on the Teaching of Mathematics," *The Monist* 25 (1915): 495–530; Florian Cajori, "The Work of William Oughtred," *The Monist* 25 (1915): 441–66; and Florian Cajori, "A List of Oughtred's Mathematical Symbols, with Historical Notes," *University of California Publications in Mathematics* 1, no. 8 (February 18, 1920).

42. See below for Oughtred's derogatory comments on instrument makers as mere "jugglers" who are concerned only with "the surface."

43. For a discussion of the prevalence of mathematics in Gresham College and the universities, see Mordechai Feingold, *The Mathematicians' Apprenticeship* (Cambridge: Cambridge University Press, 1984).

44. William Oughtred, *The Key of Mathematicks New Forged and Filed* (London, 1647). The English preface follows closely the Latin version of 1631.

45. Oughtred, preface to *The Key of Mathematicks*.

46. Ibid.

47. Ibid.

48. This passage follows the previously quoted passage in the preface to the 1631 Latin edition. The translation is from Cajori, "Work of William Oughtred."

49. Stephen Rigaud, *Correspondence of Scientific Men of the Seventeenth Century* (Oxford: Oxford University Press, 1841), 65–66.

50. Rigaud, *Correspondence*, 27.

51. Taylor, *Tudor Geography*, 24.

52. Robert Recorde, *The Castle of Knowledge* (London, 1556). For more on Recorde, see Joy B. Easton, "Recorde, Robert," in the *Dictionary of Scientific Biography*; Robert Recorde, "A Tudor Euclid," *Scripta Mathematica* 27 (1966): 339–55; Robert Steele, *The Earliest English Arithmetics* (Oxford: Oxford University Press, 1922); and Francis R. Johnson and Sanford V. Larkey, "Robert Recorde's Mathematical Teaching and the Anti Aristotelian Movement," *Huntington Library Bulletin* 7 (April 1935).

53. *The Gate of Knowledge* is mentioned on p. 68 of *The Castle of Knowledge*.

54. Robert Recorde, *The Pathway to Knowledge* (London, 1551).

55. Recorde, *Whetstone of Witte*.

56. See the dedication to *Whetstone of Witte*.

57. Robert Recorde, *The Grounde of Artes* (London, 1542).

58. The poem is on the last page of the "Preface to the Reader" in *The Castle of Knowledge*.

59. Johnson and Larkey, "Recorde's Mathematical Teaching," 63.

60. "To the Gentle Reader," in *The Pathway to Knowledge*. Recorde's address was adopted by Waters in *Art of Navigation*, xxxix.

61. See the dedication in Recorde, *Whetstone of Witte*.

62. On Blundeville, see the account in Taylor, *Mathematical Practitioners*; and the Rev. Augustus Jessopp, "Blundeville, Thomas," in *Dictionary of National Biography*, vol. 5 (London: Smith, Elder & Co., 1886). See also Jean Jacquot, "Humanisme et science dans l'Angleterre Elisabethaine: L'oeuvre de Thomas Blundeville," *Revue d'Histoire des Sciences* 6 (1953): 189–202; and Antoine de Smet, "Thomas Blundeville et l'Histoire de la cartographie du XVIe Siecle," *Rivista da Universidade de Coimbra* 27 (1979): 293–301.

63. Thomas Blundeville, *His Exercises, Containing Eight Treatises* (London, 1606).

64. Ibid., 327–28. For a more detailed explanation of the Mercator projection and meridional parts tables, see the discussion of Edward Wright's work below as well as Chapter 5 below.

65. Blundeville, *Exercises*.

66. See the preface "To the Reader" in Blundeville, *Exercises*.

67. "What cause first moved the Author to write this Arithmeticke," in Blundeville, *Exercises*.

68. Thomas Blundeville, *The Theoriques of the Seven Planets, shewing all their diverse motions, and all other Accidents, called Passions, thereunto belonging* (London, 1602).

69. See the dedication to John Aspley, *Speculum Nauticum: A Looking Glass for Sea-Men* (London, 1624). The passage is also quoted in Waters, *Art of Navigation*, 405.

70. For general information on Wright, see P. J. Wallis, "Edward Wright," in

The Dictionary of Scientific Biography; as well as general accounts in Taylor, *Mathematical Practitioners;* Waters, *Art of Navigation;* and Feingold, *Mathematicians' Apprenticeship.*

71. See n. 13 above.

72. Wright, *Certaine Errors.*

73. For an account of the cartographical and navigational problems of the period, see Waters, *Art of Navigation.*

74. Taylor speculates that Mercator used some mechanical device. See Taylor, *Mathematical Practitioners,* 336.

75. "Preface," in Wright, *Certaine Errors.*

76. Ibid.

77. Ibid.

78. Ibid.

79. Ibid.

80. Wright, *Certaine Errors,* D1.

81. John Napier, *Mirifici Logarithmorum Canonis Descriptio eiusque Usus, ut etiam in Trigonometria, ut etiam in Omni Logistica Mathematica Amplissimi, Facillimi, & Expedissimi Explicatio* (Edinburgh, 1614).

82. See n. 4, above.

83. Napier, *Description,* A2.

84. John Davies of Hereford, "In praise of the never-too-much praised Worke and Author the L. of *Marchiston,*" in Napier, *Description.* The Laird of Marchiston was, of course, John Napier.

85. Davies, "In Praise."

86. The rhetoric is also in evidence in a second introductory poem to the *Description,* by Richard Lever. Lever praises the discovery of "Logarithmes" as a gift from God, that

> . . . *hidden things* sometimes *discovered* bee:
> What many *men* and *ages* could not finde,
> Is, at the last, by some *one* brought to mind.
>
> [Later in the poem Wright is described as]
> . . . He who earst our *Navigation* clear'd
> From that strange tongue to *English* it did change,
> That famous, learned, *Errors* true corrector,
> *Englands* great *Pilot, Mariners Director.*

The themes of hidden secrets brought to light and the mathematician as an enterprising explorer are clearly present in these passages. See Richard Lever, "In the iust praise of this Booke, Author, and Translator," in Napier, *Description.*

87. John Tapp, *The Pathway to Knowledge; Containing the whole Art of Arithmeticke, bot in whole numbers and fractions* (London, 1613). On John Tapp, see Taylor, *Mathematical Practitioners;* Waters, *Art of Navigation;* and Feingold, *Mathematicians' Apprenticeship.*

88. John Tapp, *The Seamans Kalendar, or an Ephemerides of the Sun, Moone, and certaine of the most notable fixed Starres* (London, 1602). According to Waters it was repeatedly updated and five editions were published by 1615. See Waters, *Art of Navigation*, 239–40.

89. "Dedication," in Tapp, *Pathway*.

90. Waters claims that the lectures referred to were given under Sir Thomas Smith's patronage at his home by Thomas Hood and later Edward Wright. Feingold suggests that the reference is in fact to the public lectures at Gresham College. See Waters, *Art of Navigation*, 320; and Feingold, *Mathematicians' Apprenticeship*, 176.

91. "Dedication," in Tapp, *Pathway*.

92. Ibid.

93. Ibid.

94. Tapp, in fact, refers to a previous book by the same title, but claims that it was translated from the Dutch! See the preface "To the Reader," in Tapp, *Pathway*.

95. "Dedication," in Tapp, *Pathway*.

96. More detailed surveys of the mathematical practitioners of the period can be found in Taylor, *The Mathematical Practitioners*; Waters, *Art of Navigation*; and Feingold, *Mathematicians' Apprenticeship*.

Chapter 4

1. Jean Jacquot, "Thomas Hariot's Reputation for Impiety," *Notes and Records of the Royal Society of London* 9 (1952): 164–87.

2. Hilary Gatti has recently claimed for Hariot the title of "Philosophical Sceptic." It should be noted, though, that her evidence is based on an interpretation of Hariot's mathematical and scientific work, rather than on any philosophical discussions "per se" in his writings. See the 1993 Thomas Harriot Lecture at Oriel College, Oxford, printed as Hilary Gatti, *The Natural Philosophy of Thomas Harriot* (Oxford: Oriel College, 1993).

3. B.L. Add. MS 6788, f. 131v.

4. The distinction was first made by the logical positivist Hans Reichenbach, but was later popularized by Sir Karl R. Popper in *The Logic of Scientific Discovery* (1935; reprint, London: Routledge, 1992).

5. On the medieval view of a closed intellectual world, see Anthony Grafton, *New Worlds Ancient Text: The Power of Tradition and the Shock of Discovery* (Cambridge, Mass.: The Belknap Press of Harvard University Press, 1992). Needless to say, this conception of knowledge came under sustained attack in the early modern period, and Hariot's statement puts him squarely on the side of the critics.

6. See Herbert Gillot, *La querelle des anciens et des modernes en France* (Paris: Librairie Ancienne H. Champion & É. Champion, 1914); and Hyppolyte Rigault, *Histoire de la querelle des anciens et des modernes* (Paris: L. Hachette et cie, 1856).

7. George Hakewill, *An Apologie or Declaration of the Power and Providence*

of God in the Government of the World, 2d ed. (Oxford, 1630), 264; Richard Harvey, *A Theological Discourse of the Lamb of God* (London, 1590); and Thomas Nashe, *Pierce Penilesse, his Supplication to the Devill* (London, 1592). Harvey and Nashe do not mention Hariot by name, but a list Hariot composed of contemporary references to himself makes it clear that he considered himself the target of their polemics. See David B. Quinn and John W. Shirley, "A Contemporary List of Harriot References," *Renaissance Quarterly* 22, no. 1 (spring 1969): 9–26.

8. For the problems of assimilating the discovery of America within an authority-dependent conception of knowledge, see Grafton, *New Worlds Ancient Texts*; as well as J. H. Elliott, *The Old World and the New 1492–1650* (Cambridge: Cambridge University Press, 1970); J. H. Elliott, "Renaissance Europe and America: A Blunted Impact?" in *First Images of America*, vol. 1, ed. Fredi Chiappelli (Berkeley and Los Angeles: University of California Press, 1976), 11–23; and Anthony Pagden, *European Encounters with the New World* (New Haven, Conn.: Yale University Press, 1993), esp. chaps. 2 and 3. William Eamon traces the historical emergence of the view of physical knowledge as the uncovering of nature's secrets (instead of logical deduction from traditional authority) and ties this development to the discovery of America. See William Eamon, *Science and the Secrets of Nature* (Princeton, N.J.: Princeton University Press, 1994), esp. chap. 8.

9. Elliott, "Blunted Impact." Anthony Grafton provides an excellent account of the effect of the discovery on early modern scholars in his *New Worlds, Ancient Texts*. Grafton generally supports Elliott's position that most writers sought ways to accommodate the New World within the confines of traditional scholarship.

10. This passage from Cochlaeus's introduction to the 1512 edition of Pomponius Mela's *Cosmographia* is quoted in Elliott, "Blunted Impact," 14. Johannes Stamler, in his *Dyalogus . . . de diversarum gencium sectis et mundi religionibus* of 1508, and Zacharius Lilius, in his *Contra antipodes* of 1496, also rejected the new discoveries. See Mary B. Campbell, *The Witness and the Other World* (Ithaca, N.Y.: Cornell University Press, 1988), 216.

11. Pagden, *European Encounters*, chaps. 1 and 2; and Anthony Pagden, "Ius et Factum: Text and Experience in the Writings of Bartolomé de Las Casas," in *New World Encounters*, ed. Stephen Greenblatt (Berkeley and Los Angeles: University of California Press, 1993), 85–100; Pagden's chapter was first published in *Representations* 33 (winter 1991): 147–62. For another account of travel writers' efforts to strike a balance between the classical tradition and their own status as eyewitnesses to new discoveries, see Campbell, *Witness and the Other World*.

12. Thomas Hariot, *A Briefe and True Report of the New Found Land of Virginia* (New York: Dover Publications, 1972).

13. Ibid., 5.

14. Ibid., 6.

15. John Davis, *The Worldes Hydrographical Description* (London, 1595; reprinted in Albert Hastings Markham, ed., *The Voyages and Works of John*

Davis the Navigator, London: The Hakluyt Society Publications 59, 1880).
Robert Hues, *Tractatum de globis et eorum usu* (London, 1592; reprinted as
The Hakluyt Society Publications 79, London, 1889).

16. For Raleigh's *Discoverie of Guiana*, see Walter Raleigh, *The Discovery of
Guiana by Sir Walter Raleigh*, ed. V. T. Harlow (London: Argonaut Press, 1928).
For Keymis's report, see Lawrence Keymis, *A Relation of the Second Voyage to
Guiana* (London, 1596; reprint, Amsterdam: Theatrum Orbis Terrarum, 1968).

17. For a full account of Raleigh's voyage of 1617, see V. T. Harlow, ed.,
Raleigh's Last Voyage (London: Argonaut Press, 1932).

18. The term "virtual witnessing" is, of course, borrowed from Steven Shapin
and Simon Schaffer, *Leviathan and the Air Pump* (Princeton, N.J.: Princeton University Press, 1985).

19. The *standard narrative of exploration* is discussed at length in Chapter 2,
where I argue that it shaped English explorers' perceptions of the geography of
America. According to this narrative, a hidden golden land lies deep in the interior, separated from the coast by mountains and other seemingly impenetrable
barriers. The enterprising explorer must locate the hidden passages that break
through the imposing barriers, find his way to the golden land, and be rewarded
with great riches.

20. E. G. R. Taylor, ed., *The Original Writings and Correspondence of the
Two Richard Hakluyts*, vol. 2 (London: The Hakluyt Society Publications 2, no.
77, 1935), doc. 47, pp. 327–38.

21. *Two Hakluyts*, vol. 2, doc. 47, p. 328.

22. Ibid., doc. 47, p. 330.

23. Ibid., doc. 47, p. 333.

24. Ibid., doc. 58, p. 374.

25. Ibid., doc. 56, p. 366.

26. Ibid., doc. 89, p. 501.

27. Ibid., doc. 47, pp. 328, 330.

28. Hariot, *Briefe and True Report*, 7–11.

29. Ibid., 18.

30. For the gendered significance of arrangement in the form of a list, see
Patricia Parker, "Rhetorics of Property: Exploration, Inventory, Blazon," in her
Literary Fat Ladies (London: Methuen, 1987), chap. 8.

31. Hariot, *Briefe and True Report*, 6.

32. Ibid., 5.

33. Ibid.

34. Ibid., 15.

35. Ibid. For another discussion of these passages, see Hans Galinsky, "Exploring the 'Exploration Report' and its Image of the Overseas World: Spanish,
French, and English Variants of a Common Form Type in Early American Literature," *Early American Literature* 12 (1977): 5–24.

36. Hariot, *Briefe and True Report*, 31.

37. The text is appended to John White's map "The arrival of the Englishemen in Virginia," in Hariot, *Briefe and True Report*. As I mention above in Chapter 2, David B. Quinn demonstrated Hariot's authorship of the comments.

38. Hatfield MS CP 42/36. The letter is reproduced in John W. Shirley, *Thomas Harriot: A Biography* (Oxford: Oxford University Press, 1983), 230–32.

39. See Chapter 2 above.

40. See n. 38 above.

41. George Chapman, "De Guiana, Carmen Epicum," in Keymis, *Second Voyage*.

42. B.L. Add. MS 6786, f. 464. On Hariot's map sketches, see Amir R. Alexander, "Lunar Maps and Coastal Outlines: Thomas Hariot's Mapping of the Moon," *Studies in the History and Philosophy of Science* 29 (1998): 345–68.

43. See Chapter 2 above.

44. For a discussion of the adoption of the metaphor of exploration by Elizabethan mathematical practitioners, see Chapter 3 above.

45. George Chapman, *The Poems of George Chapman*, ed. Phyllis B. Bartlett (Oxford: Oxford University Press, 1941), 381–84.

46. Ibid.

47. Richard Hakluyt, *The Principal Navigations Voyages, Traffiques and Discoveries of the English Nation, in Twelve Volumes* (Glasgow: MacLehose, 1904), 7:236. See discussion in Chapter 2 above.

48. *The Principal Navigations*, 7:237.

49. Chapman, "De Guiana, Carmen Epicum," in Keymis, *Second Voyage*.

50. Letter from Johannes Kepler to Thomas Hariot, October 2, 1606, in *Johannes Keplers Gesammelte Werke*, vol. 15 (Munich: C. H. Beck, 1951), 348–52.

51. Hariot to Kepler, December 2, 1606, in *Keplers Werke*, 15:365.

52. See David C. Lindberg, *Theories of Vision from Al-Kindi to Kepler* (Chicago: University of Chicago Press, 1976), 73; and Albert Lejeune, *Recherches sur la catoptrique grecque* (Brussels: Académie royale des sciences, 1957), 152–75.

53. On theories of the rainbow, see Carl B. Boyer, *The Rainbow, from Myth to Mathematics* (Princeton, N.J.: Princeton University Press, 1987). On vision, see Lindberg, *Theories of Vision*.

54. The law of sines states that for a given medium, if a is the angle of incidence and b is the refracted angle, then $sin(a)/sin(b) = C$ where C is a constant. The law was discussed by the Dutch scientist Willebrord Snell (1581–1625) around 1620 and was published by Descartes in "La Dioptrique" appended to the *Discours de la methode* (Paris, 1637).

55. On Hariot's optics, see Johannes Lohne, "Thomas Harriott (1560–1621): The Tycho Brahe of Optics," *Centaurus* 6, no. 2 (1959): 113–21; Johannes Lohne, "Kepler und Harriot, Ihre Wege zum Berchungsgesetz," in *Internationales Kepler-Symposium, Weil du Stadt, 1971* (Hildesheim: Gerstenberg, 1973); and Johannes Lohne, "Harriot, Thomas," in *The Dictionary of Scientific Biography*.

See also John W. Shirley, "An Early Determination of Snell's Law," *American Journal of Physics* 19 (1951): 507–8; John W. Shirley, *Thomas Harriot: A Biography*, 380f. Carl B. Boyer mentions Hariot as being among the first to formulate the law of sines in *The Rainbow*, 203. Shirley places Hariot's discovery in 1597, while Lohne states that Hariot was in possession of his principle by 1602. On Kepler's work on the rainbow and on vision, see Boyer, *Rainbow*; and Lindberg, *Theories of Vision*, respectively. For a discussion of the Kepler-Hariot exchange on the rainbow, see Boyer, *Rainbow*, 183–85.

56. *Keplers Werke*, 15:366.

57. On medieval theories on the cause of refraction, see David C. Lindberg, "The Cause of Refraction in Medieval Optics," *British Journal for the History of Science* 4 (1968–69): 23–38; as well as Lindberg, *Theories of Vision*. The views of the various medieval authors differed in many details, but all adhered to versions of this basic explanation.

58. *Keplers Werke*, 15:366.

59. Ibid.

60. Ibid., 15:367.

61. Ibid.

62. Ibid.

63. *Keplers Werke*, 16:32. For a discussion of Kepler's views on light, see David C. Lindberg, "The Genesis of Kepler's Theory of Light: Light Metaphysics from Plotinus to Kepler," *Osiris*, 2d series, vol. 2 (1986): 5–42. The book of "Optics" to which Kepler refers is his *Ad Vitellionem paralipomena* (Frankfurt am Main, 1604) where he explains his views on light and vision.

64. *Keplers Werke*, 16:172.

65. Ibid., 15:367–68.

66. The original text of the "Synopsis" is in B.L. Birch MS 4458 ff. 6–8. The text is reproduced and discussed in Jean Jacquot, "Thomas Harriot's Reputation for Impiety" (see n. 1 above). For a critique of Jacquot's views, see John Henry, "Thomas Harriot and Atomism: A Reappraisal," *History of Science* 20 (1982): 267–96.

67. *Keplers Werke*, 15:368.

68. Keymis, *Second Voyage*, E2v.

69. Keymis, of course, was a friend of Hariot and dedicated to him a Latin poem by George Chapman entitled "De Guiana Carmen" in the introduction to his *Second Voyage*. The poem outlines the very same narrative: Guyana is described as having "gold and gems." It is "surrounded by many mountains as if by walls," but nonetheless "has deep bounteous entrances" that allow access to the explorers. For a discussion of the poem, see Chapter 2 above.

70. The quote is from the "Praefatio" to Nathaniel Torporley, *Diclides Coelometricae seu Valuae Astronomicae Universales* (London, 1602). I have used the translation provided by Jean Jacquot in "Thomas Harriot's Reputation for Impiety," 168. It should be noted that in later references, such as in his *Corrector Analyticus*

and the "Synopsis of the Controversy of Atoms," Torporley is far less flattering to Hariot. The "Corrector" is a Sion College MS, published (with errors) by J. O. Halliwell in his *A Collection of Letters Illustrative of the Progress of Science in England* (London: R. and J. E. Taylor, 1841), 109–16. For the "Synopsis," see n. 66 above.

71. Hariot, for example, assured Kepler of his intention to publish his work on the rainbow in his letter of December 2, 1606, *Keplers Werke*, 15:368. Elsewhere, William Lower, in a letter dated February 6, 1610, urged Hariot to publish his work so that others will not take credit for his discoveries. The letter is quoted in Shirley, *Thomas Harriot: A Biography*, 400.

72. Hariot's will was first discovered and published in Henry Stevens, *Thomas Harriot and his Associates* (London: Chiswick Press, 1900). The will is reprinted, together with a detailed account of repeated efforts to publish parts of the manuscripts, in part 2 of Rosalind C. H. Tanner, "Thomas Harriot as Mathematician: A Legacy of Hearsay," *Physis* 9, no. 2 (1967): 235–47 (part I); and no. 3 (1967): 257–92 (part II). For a further account of the publication history of the papers, see Rosalind C. H. Tanner, "Nathaniel Torporley and the Harriot Manuscripts," *Annals of Science* 25 (1969): 339–49; as well as "Harriot in History," chap. 1 in Shirley, *Thomas Harriot: A Biography*.

73. Thomas Hariot, *Artis Analyticae Praxis* (London, 1631).

74. For an account of the loss and rediscovery of the Hariot manuscripts, see "Harriot in History," chap. 1 in Shirley, *Thomas Harriot: A Biography*; Johannes A. Lohne, "The Fair Fame of Thomas Harriott: Rigaud versus Baron von Zach," *Centaurus* 8 (1963): 69–84; and R. C. H. Tanner, "Thomas Harriot as Mathematician: A Legacy of Hearsay," parts I and II. The vast majority of the surviving papers can be found in the British Library and in Petworth House in Sussex. The Petworth House manuscripts are owned by the Earl of Egremont, descendant of the ninth Earl of Northumberland, Hariot's patron. In the British Library see Add. MSS 6782–6789, and in Petworth House see HMC 240/i–v, 241/i–x.

75. There have, nonetheless, been successful attempts at interpreting parts of Hariot's scattered manuscript notes: for the most detailed and sophisticated of these, see Jon V. Pepper, "Harriot's Calculation of the Meridional Parts as Logarithmic Tangents," *Archive for the History of Exact Sciences* 4 (1967–68): 359–413; Jon V. Pepper, "Harriot's Earlier Work on Mathematical Navigation: Theory and Practice," in *Thomas Harriot, Renaissance Scientist*, ed. John W. Shirley (Oxford: Clarendon Press, 1974); John D. North, "Thomas Harriot's Papers on the Calendar," in *The Light of Nature: Essays in the History and Philosophy of Science presented to A. C. Crombie*, ed. John D. North and John J. Roche (Dordrecht: M. Nijhoff, 1985), 145–74; John J. Roche, "Harriot's 'Regiment of the Sun' and its Background in Sixteenth Century Navigation," *The British Journal for the History of Science* 14, no. 48 (1981): 245–61.

76. HMC 241/iv, ff. 23–31. Related diagrams can be found scattered in the British Library collection. A manuscript copy can be found in B.L. Harley MS 6002, ff. 16v–20.

77. B.L. Harley MS 6002, ff. 21–21v. The letter is transcribed in James O. Halliwell, *Letters Illustrative of the Progress of Science*, 45.

78. Hariot to Kepler, December 2, 1606, *Keplers Werke*, 15:365.

79. See Jon V. Pepper, "Harriot's Manuscript on the Theory of Impacts," *Annals of Science* 33 (1976): 131–51; and Martin Kalmar, "Thomas Harriot's 'De Reflexione Corporum Rotundorum': An Early Solution to the Problem of Impact," *Archive for the History of Exact Sciences* 16 (1977): 201–30. A less coherent account is J. A. Lohne, "Essays on Thomas Harriot: I. Billiard Balls and the Laws of Collision," *Archive for the History of Exact Sciences* 20, no. 3 (1979): 189–204.

80. Hariot never mentions "weight," but merely posits that the magnitudes of the two balls are in a 2:3 ratio. The context of his use, nonetheless, makes it clear that he is referring to the relative weights of the balls.

81. Hariot does not explain the meaning of the term, but his general intention is clear from the argument. See Pepper, "Theory of Impacts," 138, n. 13; and Kalmar, "Hariot's 'De Reflexione,'" 204, n. 6.

82. Pepper demonstrates that by increasing the total nutus by $1/3$ Hariot is privileging the larger ball over the smaller one. The correct ratio would be $1/5$, that is, the difference between the weight divided by their sum. Kalmar points out that this would bring Hariot's predictions to accord with modern kinematical predictions.

83. Pepper, "Theory of Impacts," 141–42.

84. *Keplers Werke*, 15:367–68.

85. HMC 241/vi, f. 23. The passage is quoted with slightly varying translations in Pepper, "Theory of Impacts"; and Kalmar, "Hariot's 'De Reflexione.'" The original runs as follows:

> Ista poristica vi sua universam scientiam de reflexione corporum designant. Ac, qui illa recte intellegit, omnium aliorum casuum et totius doctrinae est quasi Magister. Atque ideo non inepte Magisteria dici possunt. Sunt etiam dignitate inter praecipua quae ad Naturalis Philosophiae Penetralia sive Mysteria conducunt intelligenda. Aristoteles, veteres et recentiores, huiusmodi problemata propununt, media quaerunt, arguunt, concludunt. Sed secundum illud Terentianum: Faciunt nae intelligendo, ut nihil intelligant.

86. *Keplers Werke*, 15:368.

Chapter 5

1. B.L. Add. MS 6788, f. 131v.

2. Euclid's *Elements* was naturally the paradigmatic case for this view of mathematics.

3. For early modern discussions of the nature of mathematical truths, see

Paolo Mancosu, *Philosophy of Mathematics and Mathematical Practice in the Seventeenth Century* (Oxford: Oxford University Press, 1996), chap. 1; as well as Paolo Mancosu, "Aristotelian Logic and Euclidean mathematics: Seventeenth Century Developments of the Quaestio de Certitudine Mathematicarum," *Studies in the History and Philosophy of Science* 23 (1992): 241–65. For more general discussions of the classical understanding of mathematical truths and the ensuing difficulties encountered by the early calculus, see Margaret E. Baron, *The Origins of the Infinitesimal Calculus* (New York: Pergamon Press, 1969); and Carl B. Boyer, *The History of the Calculus and its Conceptual Development* (New York: Dover Publications, 1959).

4. *Johannes Keplers Gesammelte Werke*, vol. 15 (Munich: C. H. Beck, 1951), 368; my emphasis.

5. B.L. Add. MS 6782, f. 363.

6. Hariot's insistence that mathematical structures should have a real existence in the world is evident in other parts of his work as well. His unpublished treatise "De Numeribus Triangularibus" (B.L. Add. MS 6782, ff. 107–46v) deals with interpolation through the calculus of finite differences and is undoubtedly one of the earliest works on the topic. As the title of the treatise suggests, Hariot did not think he was dealing with pure abstractions. The mathematical relations he was discussing composed actual physical triangles. The treatise and the manuscript papers adjoining it are filled with sketches of number series composing actual triangles, or, alternatively, triangles and pyramids composed of parts whose relations correspond to his "triangular numbers." This tract still awaits detailed study, but it is clear that here, as elsewhere in Hariot's work, mathematical relations must "act reall or existence."

7. See Chapter 3 above.

8. That is, the ratio of any two magnitudes can be expressed as a/b where a and b are integers.

9. Since the diagonal of a square with side a is $\sqrt{2}\,a$, this result is mathematically equivalent to the modern notion that $\sqrt{2}$ is an irrational number.

10. There are four paradoxes, known by the names of Achilles, The Arrow, The Stade, and The Dichotomy. Two of them, The Arrow and The Stade, seem to attack the atomistic notion that there is a limit to divisibility. Since this position essentially follows up on the problem of incommensurability, it will not concern us here. The Dichotomy and Achilles attack the notion of infinite divisibility and will be discussed below. For general discussions of the problem of incommensurability and Zeno's paradoxes, see Boyer, *History of the Calculus*; and Baron, *Origins of the Infinitesimal Calculus*; Carl B. Boyer, *A History of Mathematics*, revised by Uta C. Merzbach, 2d ed. (New York: John Wiley & Sons, 1991); and Morris Kline, *Mathematical Thought from Ancient to Modern Times* (Oxford: Oxford University Press, 1972).

11. Aristotle, *Physics*, 233a, 23–30.

12. The "Method of Exhaustion" sought to establish the correct ratio of mag-

nitudes of geometrical figures by proving that they could be neither greater nor smaller than the correct ratio.

For example, Archimedes in "The Quadrature of the Parabola" proves that a segment of a parabola is equal to four thirds of the area of a triangle with the same base and vertex. He does so by enclosing the triangle within the parabola, and then adding layers upon layers of triangles on the side of the triangle, gradually "exhausting" the area within the parabola segment.

The sum of the triangles converges on $4/3$ the area of the original triangle, or in Archimedes' terms—the sum of the triangles could be made to differ from $4/3$ by less than any assigned magnitude. Now if one assumes that the area of the parabolic segment was greater than $4/3$ the area of the original triangle by a definite magnitude, then Archimedes could show that by adding more triangles, their sum could be made to differ from the parabola segment by less than the assigned magnitude. Therefore, the parabola segment could not be greater than $4/3$ the area of the original triangle. At the same time the parabola segment could not be smaller either, since then it would be possible for the sum of the triangles to be greater than the area of the parabola segment. That would be absurd, since the triangles are all enclosed within the area of the parabola.

Since the assumptions that the parabola segment is either greater or smaller than $4/3$ the area of the original triangle lead to logical contradictions, one must conclude that it is equal to $4/3$ the area of the original triangle.

For a detailed discussion of the method of exhaustion, see Boyer, *History of the Calculus*, 52.

13. Wilbur Knorr has argued that classical mathematicians were unaffected by the philosophical debates of their times and that the method of exhaustion was simply a "refinement" of cruder earlier mathematical techniques. While Knorr's intimate familiarity with the classical sources is beyond question, I have two main problems with his position. First, in many of his examples Knorr seems to assume what he seeks to prove—namely, the insularity of mathematics. For instance, he considers the paradoxes of Zeno to be irrelevant for mathematical purposes because they deal with physical rather than geometrical difficulties. This formal distinction between physics and mathematics was not at all self-evident to philosophers and geometers of the time; it was, rather, one of the central issues at stake in the controversy over infinitesimal methods. It cannot be assumed as a universal principle and then applied to demonstrate that the mathematical tradition was self-contained. Even more problematic is Knorr's view of the development of the method of exhaustion as a natural "refinement" of earlier infinitesimal approaches. The method can only be considered an improvement over earlier practices if one assumes the Euclidean standards of rigor. Knorr here assumes that the Euclidean forms are a necessary development of mathematics and then uses this standard to argue that earlier methods leading up to Euclid are "technical improvements." Knorr, in other words, assumes what he sets out to prove. One may or may not accept his contention that the philosophical discussions make use of the work of

the mathematicians, rather than the other way around; in either case it is clear that the fundamental questions expressed in the paradoxes of Zeno and the problem of incommensurability played a crucial role in development of the method of exhaustion and in classical mathematics' general avoidance of infinitesimal techniques. See Wilbur R. Knorr, "Infinity and Continuity: The Interaction of Mathematics and Philosophy in Antiquity," in *Infinity and Continuity in Ancient and Medieval Thought*, ed. Norman Kretzmann (Ithaca, N.Y.: Cornell University Press, 1982).

14. Archimedes' *Method*, discovered by J. L. Heiberg in 1899, clearly shows that he was quite satisfied with the reliability and accuracy of infinitesimal methods for his own purposes. For the best account of Archimedes' work, see E. J. Dijksterhuis, *Archimedes* (1938; reprint, Princeton, N.J.: Princeton University Press, 1987).

15. The method of exhaustion avoided the use of infinitesimals by proving that the required magnitude lies between two magnitudes that approach each other in the limit to any degree required. The area of a circle, for example, lies between the areas of an enclosed polygon and a similar enclosing polygon. When the number of sides of the two polygons are increased, their areas approach each other to any degree required. It is then shown by "reductio ad absurdum" that the circle that is blocked between the two polygons can be neither greater nor lesser than the area that both polygons approach. For more detailed accounts and examples of this method, see Boyer, *History of the Calculus*; Boyer and Merzbach, *History of Mathematics*; Baron, *The Origins of the Calculus*; and Kline, *Mathematical Thought*.

16. Simon Stevin, *De Bghinselen der Weeghconst* (Leyden, 1586).

17. For Stevin's method, see E. J. Dijksterhuis, *Simon Stevin, Science in the Netherlands around 1600* (The Hague: Nijhoff, 1970), 57–59.

18. Johannes Kepler, *Nova stereometria doliorum vinariorum* (Linz, 1615). See discussion in Boyer, *History of the Calculus*, 107f.

19. Galileo Galilei, *Two New Sciences*, ed. and trans. Stillman Drake (Madison: University of Wisconsin Press, 1974), 27f.

20. Bonaventura Cavalieri, *Geometria indivisibilibus continuorum nova quadam ratione promota* (Bologna, 1635); and Bonaventura Cavalieri, *Exercitationes geometricae sex* (Bologna, 1647).

21. Cavalieri, *Exercitationes*, 3.

22. Paul Guldin, *Centrobaryca seu de centro gravitatis trium specierum quantitatis continuae*, 4 vols. (Vienna, 1635–41).

23. This curve is commonly known today as a logarithmic spiral and is defined by the equation $r = e^\theta$.

24. The equiangular spiral belongs to the class of "mechanical curves" ("transcendental curves" in modern terms) that Descartes wanted to exclude from geometry. In *La géométrie* of 1637 Descartes claimed that such curves were too vague and inexact to be accurately analyzed. In spite of Descartes's warning, in 1645 Evangelista Torricelli published his detailed analysis of the properties of the

equiangular spiral. Hariot had already obtained these results thirty years before, but never published them. Thus, twenty-four years after Hariot's death, "his" spiral became known as the first curve to be rectified in modern times. For Torricelli's work, see Evangelista Torricelli, *De Infinitis Spiralibus*, ed. Ettore Carruccio (Pisa: Domus Galilaeana, 1955). On Descartes's and Torricelli's views on curves, see Boyer, *History of Mathematics*, 341–43.

25. For more on the navigational problem and its solutions, see Chapter 3 above; as well as David W. Waters, *The Art of Navigation in England in Elizabethan and Early Stuart Times* (New Haven, Conn.: Yale University Press, 1958); A. H. W. Robinson, *Marine Cartography in Britain* (Leicester: Leicester University Press, 1962); and E. G. R. Taylor, *Tudor Geography* (London: Methuen, 1930).

26. Mercator is compensating for one distortion by creating another. Rather than depict the longitude degrees as shrinking while keeping the latitudes fixed (as is the case on the globe), he keeps the longitude degrees fixed while gradually increasing the degrees of latitude. As a result, at any given point on the chart the ratio between a latitude and longitude degree is correct.

27. E. G. R. Taylor, "John Dee and the Nautical Triangle," *Journal of the Institute of Navigation* 8 (October 1955): 318–25.

28. Edward Wright, *Certaine Errors in Navigation* (London, 1599).

29. The secant of an angle is the inverse of its cosine, that is, $sec(a) = 1/cos(a)$.

30. All figures here are from Wright, *Certaine Errors*. It should be noted that in this projection, the length of degrees of latitude approaching the poles tends to infinity. Although Wright uses the "addition of secants" method in calculating his tables, he nonetheless provides a somewhat different explanation of his proceedings. He envisions a hollow cylinder, enveloping the globe while touching it only at the equator. He now suggests that the globe be expanded, like a balloon, through the hypothetical cylinder. Each circle of latitude will be marked on the cylinder at the height where the given latitude on the expanding globe intersects with the cylinder. The projection is equivalent to the addition of secants. Florian Cajori, writing in 1914, was satisfied to call Wright's addition procedure an "integration." See Florian Cajori, "On an Integration ante-dating the Integral Calculus," *Bibliotheca Mathematica* 14 (1914–15): 312–19.

31. Edward Wright, "A Cruizing Voyage to the Azores in 1589 by the Earl of Cumberland," in *A General History and Collection of Voyages and Travels*, vol. 7, ed. Robert Kerr (Edinburgh: W. Blackwood, 1824), 375–96. This text was appended to the early editions of *Certaine Errors*. For other references to Wright as an enterprising explorer, see Chapter 3 above.

32. Jon V. Pepper, "Harriot's Earlier Work on Mathematical Navigation: Theory and Practice," in *Thomas Harriot, Renaissance Scientist*, ed. John W. Shirley (Oxford: Oxford University Press, 1974), 54–83. It should be noted that in the introduction to *Certaine Errors* Wright says that his manuscript was circulating for several years before its publication in 1599. It is quite possible that Hariot had seen it during that time. For an earlier account of Hariot's work on navigation, see

"Appendix No. 30: Thomas Hariot's Contributions to the Art of Navigation," in Waters, *Art of Navigation*. Clements R. Markham suggests that Hariot was the author of section 5 of Robert Hues's *Tractatus de Globis* of 1594, which deals with rhumbs. I have found no textual support for this assertion, and in any case, the section does not discuss the problem of projection. See Robert Hues, *Tractatus de Globis*, ed. Clement R. Markham (London: The Hakluyt Society Publications, no. 79, 1889).

33. A "radius vector" here is the line connecting a given point on the "rhumb" with the pole.

34. This account of Hariot's work on the Mercator projection and much of the technical material discussed below are based on Jon V. Pepper, "Harriot's Calculation of the Meridional Parts as Logarithmic Tangents," *Archive for the History of Exact Sciences* 4 (1967–68): 359–413. A less detailed account by the same author can be found in Jon V. Pepper, "Harriot's Unpublished Papers," *History of Science* 6 (1968): 17–40; and a response to critics is contained in Jon V. Pepper, "Some Clarifications of Harriot's Solution of Mercator's Problem," *History of Science* 14 (1976): 235–44. Hariot's study of the continuum is contained in his unpublished tract "De Infinitis," B.L. Add. MS 6782, ff. 362–74v, as well as in many folios on the topic scattered throughout his papers. These texts will be discussed in detail below.

35. Quoted in Robert Hues, *Tractatus de Globis*, 133. Hues (or Hariot, according to Markham) takes issue with Frisius's claim that the rhumbs actually meet at the poles.

36. John Davis, "The Seamans Secrets," in *The Voyages and Works of John Davis the Navigator*, ed. Albert Hastings Markham (London: The Hakluyt Society Publications, no. 59, 1880), 315.

37. For a more technical summary of Hariot's work on the Mercator projection, see the Mathematical Appendix below. A detailed technical account can be found in Pepper, "Harriot's Calculation."

38. Hariot, in fact, did not settle for the β minutes table, but rather produced separate tables for whole degrees and 10ths of minutes as well. When comparing his tables with the values of $\tan(\pi/4 - \phi/2)r_0$ he started out with the "degree table," moved to the "minutes table" and eventually to the "10th of a minute" table, thus determining the meridional parts more efficiently. See Pepper, "Harriot's Calculation." Pepper also describes the sophisticated interpolation techniques used by Hariot in constructing the table for 10th of a minute.

39. Petworth House collection, HMC 240, ff. 211–53. See discussion in Pepper, "Harriot's Calculation," 370f.

40. See Pepper, "Harriot's Calculation," 371–72. Pepper also clarifies the difficulties involved in this process. Summing up an infinite geometric series was itself a novel process that occupied Hariot a great deal. See discussion below on his treatise "De Infinitis."

41. As Pepper points out, in modern terms one would say that these calcula-

tions involve a double-limiting process. First, for a given regular angular interval θ, the sum of the radius- vectors/sides/areas is calculated as the sum of an infinite series. Hariot then repeats the process for a smaller and smaller θ, thus providing a closer approximation of the true spiral. He then deduces the limiting value of those sums as θ tends to 0.

42. The two problems correspond to the two phases of Hariot's "double-limiting procedure." First, Hariot calculates the length of the approximate spiral, composed of similar triangles with central angle θ. Then he calculates the length of the true spiral by letting θ approach 0. In doing so he is assuming that the smooth spiral is composed of infinitesimal triangles.

43. In one of his manuscript papers Hariot seems to be meditating on a different issue involved in analysis of the rhumb—the problem of incommensurability. In B.L. Add. MS 6786, f. 402, he writes, "That a magnitude cannot be divided into parts of all proportions by no numbers rationall nor no number if irrationall simple nor composed if the composition be finite."

The impression that Hariot is here indeed meditating on his analysis of the spiral is strengthened when we note that in an adjacent page, f. 400, Hariot sketches the familiar outlines of the equiangular spiral and its approximation through a geometric series of similar triangles.

44. For Hariot's speculations on the "Achilles" paradox, see B.L. Add. MS 6784 ff. 246–48, 359, 360, Add. MS 6785 f. 437, Add. MS 6786 f. 349v, 472, Add. MS 6789 ff. 103r+v.

45. Any other ratio of speeds besides 1:10 would, of course, do just as well.

46. B.L. Add. MS 6784, ff. 246–48. F. 430 from the same manuscript volume also clearly belongs to this set. The signs used in f. 246 undoubtedly refer to the sketches of the triangles in Add. MS 6785, f. 437, which appear under the single heading "Achilles."

47. B.L. Add. MS 6789, f. 103.

48. B.L. Add. MS 6786, f. 349v.

49. That is, "The Gordian Knot, The Labyrinth of Daedalus." B.L. Add. MS 6786, f. 349v.

50. See Chapter 3 above.

51. Francis Bacon, *The New Organon and Related Writings* (London, 1620; reprint, New York: Liberal Arts Press, 1960), 12.

52. William Oughtred, *The Key of Mathematicks New Forged and Filed* (London, 1647). The preface follows closely the Latin of the first edition of 1631.

53. See preface to Edward Wright, *Certaine Errors*.

54. "The Club of Hercules: Democritus his reason for atoms, Achilles: Zeno's reason." See B.L. Add. MS 6786, f. 349v.

55. Hariot borrowed the term "Club of Hercules" from Roger Bacon. In refuting atomism in chapter 39 of his *Opus Tertium*, Bacon refers to Leucipus' and Democritus' main argument for atomism as "Clava Herculis." See discussion below.

56. In the section of "De Infinitis" entitled "Ratio Achilles," B.L. Add. MS 6782, f. 367, Hariot clearly states that the Achilles paradox contradicts Aristotle's doctrine of infinite divisibility. In other words, according to Hariot, the Achilles supports mathematical atomism. See discussion below.

57. The main body of the tract "De Infinitis" is in B.L. Add. MS 6782, ff. 362–74v, although single folios entitled "De Infinitis" are scattered throughout Hariot's papers. "Ratio Clava Herculis" is the title Hariot gave to f. 372, and "Ratio Achilles" is the title he gave to folios 367 and 368, both of B.L. Add. MS 6782. The "De Infinitis" is a complex tract, and has been the subject of a great deal of scholarly attention: Jean Jacquot first drew attention to it in "Thomas Hariot's Reputation for Impiety," *Notes and Records of the Royal Society of London* 9 (1952): 164–87, where he concluded that Hariot favored mathematical atomism. Robert Kargon supports Jacquot's conclusions in his *Atomism in England from Harriot to Newton* (Oxford: Clarendon Press, 1966); and Hillary Gatti draws similar conclusions most recently in her published lecture *The Natural Philosophy of Thomas Harriot* (Oxford: Oriel College, 1993). John Henry, by contrast, is skeptical of Hariot's atomism in his article "Thomas Harriot and Atomism: A Reappraisal," *History of Science* 20 (1982): 267–96. My own interpretation strongly supports the views of Jacquot, Kargon, and Gatti, which are upheld by the evidence I present below.

58. The connection between Hariot's discussion of rhumbs and his general investigation of the continuum is also manifested in certain passages in the "De Infinitis," which are directly inspired by Hariot's work on the spiral. One example is Hariot's discussion in B.L. Add. MS 6782, f. 365 of how "in a finite time an infinite space may be moved." See discussion below. Another example can be found in f. 371, where Hariot questions the possibility "that a finite line may have an infinite number of parts. & if all the parts be in continuall proportion: The number must be compounded of an infinite number of finite partes." This is, in essence, a restatement of the paradox of the rhumb in more general terms.

59. For more on Aristotle's views on the continuum and his ancient, medieval, and early modern commentators, see John E. Murdoch, "Naissance et developement de l'atomisme au bas moyen age Latin," in *La science de la nature: theories et pratique*, Cahiers d'etudes medievales 2 (Montreal: Bellarmin, 1974); John E. Murdoch, "William of Ockham and the Logic of Infinity and Continuity," in *Infinity and Continuity in Ancient and Medieval Thought*, ed. Norman Kretzmann (Ithaca, N.Y.: Cornell University Press, 1982); John E. Murdoch, "Infinity and Continuity," in *The Cambridge History of Later Medieval Philosophy*, ed. Norman Kretzmann, Anthony Kenny, and Jan Pinborg (Cambridge: Cambridge University Press, 1982), 564–91; and John D. North, "Finite and Otherwise. Aristotle and Some Seventeenth Century Views," in *Nature Mathematized*, ed. William R. Shea (Dordrecht: D. Reidel, 1983).

60. B.L. Add. MS 6782 f. 362. The entire text as Hariot wrote it reads:

Aristotle in the beginning of his 6th booke of his Phisickes, & in the 26th text of the 5th booke, defined those things to be Continuo quorum extrema sunt unum. And in the 22th text of the sayd 5th booke that: Tangentia sunt quorum extrema sunt simul. Simul, qui in uno loco sunt primo. Separatim, qui sunt in altero. Now for the better explication of the meaning of the definitions as also of there truth, let us understand first two materiall cubes A&B to be separate, that is to be in diverse places extremes and all.

For a full translation of the *Physics*, see Aristotle, *The Works of Aristotle Translated into English*, vol. 2, ed. W. D. Ross (Oxford: Clarendon Press, 1930). Hariot's place references throughout his notes correspond to the edition of Julius Pacius, *Aristotelis Naturalis Ausculationis* (Frankfurt, 1596). I have not found this numeration in any other edition of Aristotle.

61. Aristotle's discussion can be found in the beginning of book 6 of the *Physics*. See Aristotle, *Physics*, 231a, 20–231b, 6. For a detailed interpretation, see David Bostock, "Aristotle on Continuity in *Physics* VI," in *Aristotle's Physics: A Collection of Essays*, ed. Lindsay Judson (Oxford: Clarendon Press, 1991), 179–212.

62. Thomas Hariot to Johannes Kepler, July 13, 1608, in *Johannes Keplers Gesammelte Werke*, vol. 16 (Munich: C. H. Beck, 1954), 172.

63. B.L. Add. MS 6782, f. 367. In the following page, f. 368, Hariot proceeded to elaborate the exact courses of Achilles and the tortoise and their relation to each other.

64. Aristotle, *Physics*, 239b, 25–29. Translation is from W. D. Ross, ed., *Works of Aristotle*.

65. Aristotle, *Physics*, 233a, 23–30.

66. This seems to be Hariot's meaning in B.L. Add. MS 6782, f. 374, where he writes:

> Of contradictions that spring from diverse suppositions it cannot truly be sayd that the one parte or other is false, for they are true consequences from the suppositions & in that respect are both true. but that which followeth, is, that one of the suppositions is necessaryly false, from whence one of the partes of the contradiction was inferred. As in the reason Achilles & other reasons of Zeno &c.

Since the Achilles assumed infinite divisibility and arrived at a contradiction, this assumption must be false. See also Jacquot, "Hariot's Reputation," 178.

67. See the sketches on B.L. Add. MS 6782 f. 370v, which, like those on f. 362, show the continuum as composed of distinct sections joined together.

68. Aristotle's discussion can be found in *De Caelo* I.5, 272b. 25–29. The fact that Hariot draws the paradox from Aristotle's discussion (and rejection) of an infinite universe may prove relevant to the ongoing debate about Hariot's relationship to Giordano Bruno. Jean Jacquot, Robert Kargon, and Hilary Gatti view Bruno as a major influence on Hariot, while John Henry rejects this connection (for references see n. 57 above). Since Bruno was the most prominent advocate

of an infinite universe, Hariot's evident interest in the subject seems to support the notion of a connection with Bruno.

69. B.L. Add. MS 6782, f. 365. It should be noted that the stated subject of this discussion is a direct response to Hariot's concerns about rhumbs. In the note on the paradox of the rhumb (B.L. Add. MS 6786, f. 349v) Hariot questions how it is possible to traverse an infinite number segments in a finite time. His study of how "in a finite time an infinite space may be moved" is clearly related.

70. Aristotle, *De Caelo* I.5, 272b. 29.

71. Hariot discusses the paradoxes derived from the existence of this finite angle in the following page, f. 366. I will not elaborate these questions here.

72. As noted, the quote is from B.L. Add. MS 6782, f. 365.

73. It is difficult to determine what Hariot means by "implicat." The same term is used a few lines earlier, where it refers to the argument that a line cannot be in two places at once. Since Hariot is here essentially repeating the same argument (for infinite rather than finite lines), we can conclude that in using the term Hariot is reintroducing the very same objection in this case as well.

74. Duns Scotus (ca. 1265–1308), *Opus Oxoniense*, Lib. II, dist. 2, quest. 9. Duns Scotus became the standard reference for this paradox, although it is likely that he did not author it himself.

75. The Latin *deinceps* translates roughly as "in succession." The phrase, then, means "the atoms which are the ones next to the point *a*."

76. The entire passage is in B.L. Add. MS 6782, f. 369.

77. Scotus, *Opus Oxoniense*, Lib. II, dist. 2, quest. 9.

78. B.L. Add. MS 6782, f. 369.

79. See, for example, Hariot's discussion of the "Achilles" paradox in B.L. Add. MS 6784, f. 359, his discussion, based on the *Physics*, 233b, 16–32, of a paradox similar to Zeno's "Stade" in B.L. Add. MS 6784, f. 429, the "Ratio aequalitatis in infinitum" in B.L. Add. MS 6786, f. 472, the discussion entitled "Ex Linea Quadratura Producta, consequences quidam miranda" in B.L. Add. MS 6785, f. 190, and the set of questions entitled "De Infinitis" in B.L. Add. MS 6785, f. 436. The Petworth House collection also contains several experiments with infinite summations, especially HMC 241. 1ff. 105–7. B.L. Add. MS 6785, f. 190v undoubtedly belongs with these Petworth House manuscripts.

80. B.L. Add. MS 6782, f. 374.

81. B.L. Add. MS 6782, f. 369.

82. B.L. Add. MS 6786, f. 349v. "Clava Herculis. Democritus, his reason pro atomis."

83. Roger Bacon, "Opus Tertium," in *Rerum Britannicorum Medii Aevi Scriptores (The Rolls Series)*, ed. J. S. Brewer, vol. 15 (London: H. M. Stationary Office, 1859), chap. 39, 131–35. Bacon's source for Democritus' argument was apparently Aristotle's discussion of the continuum in book 1 of *De Generatione*. Bacon cites Aristotle's responses to Democritus but finds them lacking, whereupon he introduces his own argument against atomism. Aristotle's discussion of Dem-

ocritus' argument can be found in Aristotle, *De Generatione et Corruptione*, ed.
C. J. F. Williams (Oxford: Clarendon, 1982), book 1, chap. 2. Bacon's source for
Democritus' argument is probably on p. 7 (I.2.316b, 19–34).

84. Gatti shows that this passage is most likely inspired by a similar one
in Bruno's *De Triplici Minimo*. See Hilary Gatti, *The Renaissance Drama of
Knowledge* (New York: Routledge, 1989), 60–61. Her claim in the 1993 Thomas
Harriot Lecture at Oriel College, Oxford, that Hariot's text is a "translation" of
Bruno seems to me exaggerated. See Gatti, *Natural Philosophy*, 2. In the *Drama
of Knowledge* Gatti makes a strong case for the Northumberland circle's connection to Bruno and his writing. She argues that Hariot's preoccupation with mathematical atomism, and the above-quoted passage in particular, closely follow
Bruno's line of reasoning. I am sympathetic to Gatti's interpretation and her
emphasis on the dramatic narrative of knowledge in the Renaissance. There is
no doubt that "Giordano Bruno's drama of knowledge," which she quotes from
Bruno's *Cena della Ceneri* on p. 32, bears a strong resemblance to the Elizabethan
exploration narrative. My own concern, however, is not with the Continental
influence, but rather with the local English imperialist context in which the
narrative of discovery dominated.

85. B.L. Add. MS 6782, f. 363.

86. Ibid.

87. Ibid.

88. It appears that for Hariot the existence of the "universally infinite" followed necessarily from the existence of indivisibles, and vice versa. For if a magnitude is composed of indivisibles, then there is a definite number of those in a given
magnitude, which must be the "universally infinite." Conversely, if there is a maximum "universally infinite" number, then the division of a magnitude must stop at
that number, thus dividing the continuum into indivisibles.

89. The objection to "mere suppositions" is probably a reference to Aristotle's
doctrine of the "potential infinite." In book III of the *Physics* Aristotle distin-
guishes between "potential" and "actual" infinite divisibility of the continuum.
Although it is potentially possible, according to Aristotle, to infinitely divide a
magnitude, it is not "actually" possible to do so. Hariot denies the distinction
and insists that infinite divisibility results in "actual" infinitesimals.

90. B.L. Add. MS 6782, f. 374v. In the original Latin the passage reads:

Tangentia sunt existentia in duobus locis per individuam distantiam.
Continuum est aggregatum Tangentium
Minimum continuum est aggregatum duarum tangentium.
Continuum primo componitur ex individuis tangentibus.

91. B.L. Add. MS 6782, f. 374.

92. In his discussion of the "paradox of the rhumb," where Hariot refers to
"Achilles" and "Clava Herculis," he also refers to "Abyssus, Cerberus, Tartarae,
Cimmeriae Tenebrae." See B.L. Add. MS 6786, f. 349v.

93. Hariot also recruits Atlas to his quest, but I found his notes too obscure to be used in this context. See B.L. Add. MS 6785, f. 192. For a discussion, see Gatti, *Natural Philosophy*, 15.

94. The terminology used by Hariot in all three instances is also remarkably similar, as evidenced by the Latin edition of his text appended to John White's map "The Arrival of the Englishemen in Virginia" in the de Bry edition of Hariot's *Briefe and True Report*. The text itself is a classic example of the use of the standard narrative of exploration and is discussed in Chapters 2 and 4 above. In the Latin version, Hariot refers to the arrangement of the islands protecting the Virginia coast as "discrete," which is precisely the term he uses to describe the pattern of indivisibles in the continuum in "De Infinitis" (B.L. Add. MS 6782, f. 370). When describing the difficulty of navigating the big ships through the passages, he claims that this was caused "ab angustias" (by their narrowness). These are precisely the words he used in his correspondence, when he asked Kepler whether he was having trouble penetrating the doors of nature's mansion "propter illarum angustias." In all instances Hariot is repeating a similar tale, using a similar terminology.

95. On Hariot's sketches of coastal maps, see Amir R. Alexander, "Lunar Maps and Coastal Outlines: Thomas Hariot's Mapping of the Moon," *Studies in the History and Philosophy of Science* 29 (1998): 345–68. The sketches can be found in B.L. Add. MS 6782, f. 1v; B.L. Add. MS 6786, f. 464; B.L. Add. MS 6787, f. 473v; and B.L. Add. MS 6787, f. 505v.

96. See Chapter 4 above.

Chapter 6

1. Giordano Bruno, *Ash Wednesday Supper*, trans. Stanley L. Jaki (The Hague: Mouton, 1975), 55. First published as *Cena de le Ceneri* (London, 1584).

2. Ibid., 56.

3. Ibid., 59. The quote is from Seneca's *Medea*, lines 201–2.

4. Bruno, *Ash Wednesday Supper*, 59. The quote is from Seneca's *Medea*, lines 375–79. For more on Bruno's laudatory references to Columbus, see Paula Findlen, "Il nuovo Colombo: Conoscenza e ignoto nell'Europa del Rinascimento," in *La rappresentazione dell'altro nei testi del Rinascimento*, ed. Sergio Zatti (Lucca: M. Pazzini Fazzi, 1998), 219–44.

5. Bruno, *Ash Wednesday Supper*, 61.

6. On Boyle's suspicion of mathematical studies, see Steven Shapin, *A Social History of Truth* (Chicago: University of Chicago Press, 1994), chap. 7; and Steven Shapin, "Robert Boyle on Mathematics: Reality, Representation, and Experimental Practices," *Science in Context* 2 (1988): 23–58. On rhetoric of discovery, see William Eamon, *Science and the Secrets of Nature* (Princeton, N.J.: Princeton University Press, 1994), chap. 8, esp. his discussion of Glanvill, 273; and Findlen, "Il nuovo Colombo."

7. On the importance of exploration rhetoric in early modern science, see Eamon, *Science and the Secrets of Nature*; and Findlen, "Nuovo Colombo." Elsewhere Eamon discusses the emergence of the imagery of science as a "hunt" in the sixteenth and seventeenth centuries and its close relationship to the imagery of exploration. Eamon demonstrates how the new imagery helped transform science from the demonstration of the known to a search for hidden and elusive secrets of nature. See William Eamon, "Science as a Hunt," *Physis*, new series vol. 31 (1994): 393–432.

Many early modern scientists did, of course, embrace mathematics as a powerful tool in the investigation of nature. My central claim is that the relationship of mathematics to the experimental sciences was highly problematic and that the attempt to integrate them fundamentally changed the nature of mathematics.

8. In the introduction to his biography of Pierre de Fermat, Michael Mahoney identifies no less than five different types of professions and activities that could be defined as "mathematics" in this period. I am referring here to the official academic study that he labels "classical mathematics." See Michael S. Mahoney, *The Mathematical Career of Pierre de Fermat* (Princeton, N.J.: Princeton University Press, 1973), chap. 1.

9. Christopher Clavius, "In disciplinas mathematicas prolegomena," in his *Opera Mathematica*, vol. 1 (Mainz, 1611), 3; quoted in Peter Dear, *Discipline and Experience: The Mathematical Way in the Scientific Revolution* (Chicago: University of Chicago Press, 1995), 40. It should be noted that Clavius himself did not always insist on this account of the nature of mathematics. More than twenty years after the "Prolegomena," in the introduction to his *Algebra* of 1609 Clavius presented a different view of this mathematical field. The purpose of algebra, he wrote, is to extract certainty from sense knowledge and to "explore a secret quantity." He then goes on to compare algebra to a hunting dog that tenaciously tracks down its prey until it lays it at its master's feet. See Christopher Clavius, "Proemium de Algebrae Praestantia," in his *Algebra Christophori Clavii Bambergensis* (Geneva, 1609).

This imagery is, of course, much closer to the exploration rhetoric than his programmatic statement, which takes Euclidean deduction as its model for proper mathematical procedure. Significantly, Clavius uses this imagery to describe algebra, another emerging branch of the new mathematics. The rhetoric of exploration was used in connection with various aspects of the new mathematics, ranging from mathematical notation to analysis. The focus of this article will be on the role of the imagery of exploration in shaping and promoting infinitesimal techniques.

10. Quoted in James A. Lattis, *Between Copernicus and Galileo: Christoph Clavius and the Collapse of Ptolemaic Astronomy* (Chicago: University of Chicago Press, 1994), 35.

11. Clavius's position is representative of the group Mahoney refers to as "classical mathematicians." As Mahoney makes clear, other groups of mathematical practitioners were at the time pursuing more practical uses for mathematics

while making far more modest claims (if any) as to the status of mathematical knowledge. See Mahoney, *Fermat*, chap. 1. On contemporary views on mathematics, see Paolo Mancosu, *Philosophy of Mathematics and Mathematical Practice in the Seventeenth Century* (Oxford: Oxford University Press, 1996); Dear, *Discipline and Experience*, esp. chap. 2; Mordechai Feingold, *The Mathematicians' Apprenticeship: Science and Society in England, 1560–1640* (Cambridge: Cambridge University Press, 1984); and Mahoney, *Fermat*, chap. 1.

12. This critique of mathematics was leveled most specifically against the practice that Mahoney refers to as "Classical Mathematics."

13. Francis Bacon, "Of the Dignity and Advancement of Learning" (Translation of *De Augmentis Scientarum*), chapter 6, in *The Works of Francis Bacon*, ed. James Spedding, Robert Leslie Ellis, and Douglas Denon Heath (London: Longman, 1860), 4:370.

14. Bruno, *Ash Wednesday Supper*, 55.

15. Many early modern scientists did, of course, embrace mathematics as a powerful tool in the investigation of nature. My claim is that the attempt to adapt mathematics to the emerging experimental sciences was a highly problematic venture. The attempt to fit mathematics into an experimental mold fundamentally changed the nature of the field.

16. "In these centuries," Paolo Rossi has noted, "there was continuous discussion with insistence that bordered on monotony about a logic of discovery conceived as a *venatio*, a hunt—as an attempt to penetrate territories never known or explored before." See Paolo Rossi, *Philosophy, Technology, and the Arts in the Early Modern Era*, trans. Salvator Attanasio (New York: Harper & Row, 1970), 42. On the relationship between the theme of *venatio* and the rhetoric of exploration, see n. 52 below.

17. It should be emphasized that the view of mathematicians as explorers did not exclude admiration for the ancients but most often went hand in hand with it. Many early modern mathematicians claimed that some of their ancient predecessors, most notably Archimedes, had indeed practiced mathematics as an exploration of the unknown, but in their formal presentations chose to erase all traces of their voyage. Modern mathematicians must now recover the old "methods of discovery" in order to further expand "the mathematical empire." See n. 5 above.

18. I am not claiming that rhetoric of exploration was used by promoters of infinitesimals to the exclusion of all others. Just as it would be silly to claim that only followers of the strict Baconian program utilized this language in natural philosophy, it would be equally wrong to argue that the theme of "exploration" was monopolized exclusively by infinitesimalists. The theme was used in different ways by different authors. Nevertheless it helped frame new kinds of questions, as well as acceptable answers, in a manner that was well suited for infinitesimal investigations.

19. Simon Stevin, *The Principal Works of Simon Stevin*, 6 vols., ed. D. J. Struik, vol. 2a (Amsterdam: C. V. Swets & Zeitlinger, 1958), 137.

20. Ibid., 2a:392. Stevin's characterization of decimal notation as a "discovery" comparable to finding an "unknown island" may seem surprising to a modern reader, used to regarding it as a notational technique rather than an uncovering of something "out there" in the world. The important point for the argument is that Stevin himself viewed it as a "discovery" and described its invention in terms of a geographical voyage of exploration.

21. The term *inventio* in Latin covers both our modern terms "invention' and "discovery."

22. I thank Peter Dear for this point.

23. The movement here is again similar to what was occurring in natural philosophy. It is often noted that practicing artisans possessed a longstanding empirical tradition of knowledge, which was separate from the official learned tradition and ignored by it. Bacon and his fellow reformers drew on this tradition and transformed it into legitimate scholarly knowledge. In the same manner it is less than surprising to find the view that mathematics is a voyage of exploration voiced by a practicing engineer, before it is later legitimized by Galileo, Torricelli, Wallis, and their colleagues.

24. See, for example, 186–88 in Galileo Galilei, *Dialogue Concerning the Two Chief World Systems*, trans. and ed. Stillman Drake (Florence, 1632; reprint, Berkeley and Los Angeles: University of California Press, 1967).

25. In the *Two New Sciences* of 1638, for example, Galileo noted "conclusions that are true may seem improbable at first glance, and yet when only some small thing is pointed out, they cast off their concealing cloaks and, thus naked and simple, gladly show off their secrets." See Galileo Galilei, *Two New Sciences*, trans. and ed. Stillman Drake (Madison: University of Wisconsin Press, 1974), 14.

26. Galileo, *Two World Systems*, 104.

27. Galileo to Torricelli, September 27, 1641, letter 15 in Evangelista Torricelli, *Opere di Evangelista Torricelli*, 4 vols., ed. Gino Loria and Giuseppe Vassura (Faenza: Stabilimento Typo-Litografico G. Montanari, 1919), 3:60.

28. In identifying a narrative of exploration and discovery I look for three standard elements. First, great riches and marvels must be posited in a hidden land; second, the land must be protected by natural barriers, such as mountains, forests, or mists; and third, clear passageways to the land are then opened by the intrepid explorer. This simple story was used by explorers from Columbus and Magellan to Frobisher and Raleigh. All the basic elements are present in the correspondence of Galileo, Cavalieri, and Torricelli. For more on the narrative of exploration, see Amir R. Alexander, "The Imperialist Space of Elizabethan Mathematics," *Studies in the History and Philosophy of Science* 27 (1995): 559–91.

29. Evangelista Torricelli, *Opere Scelte di Evangelista Torricelli*, ed. Lanfranco Belloni (Turin: Unione Tipografico Editrice Torinese, 1975), 624.

30. Cavalieri's reputation rests on his two books, in which he sought to systematically and rigorously elaborate infinitesimal methods. See Bonaventura

Cavalieri, *Geometria indivisibilibus continuorum nova quadam ratione promota* (Bologna, 1635); and Bonaventura Cavalieri, *Exercitationes geometricae sex* (Bologna, 1647).

31. Torricelli, *Opere Scelte*, 383.

32. Ibid., 382.

33. I thank Peter Dear and Michael Mahoney for pointing out the classical reference in the "royal road" metaphor. Significantly, Torricelli uses Euclid's imagery to point out Euclid's error: Cavalieri's method of indivisibles is the long-sought "royal road."

34. Bonaventura Cavalieri to Evangelista Torricelli, August 20, 1641, letter no. 12 in *Opere di Evangelista Torricelli*, 3:57.

35. Bonaventura Cavalieri to Evangelista Torricelli, May 12, 1643, letter no. 53 in *Opere di Evangelista Torricelli*, 3:123.

36. *Opere di Evangelista Torricelli*, vol. 3, letters no. 14, 20, 36, 39, 74, 82.

37. The trope of hidden gems of knowledge and secret natural marvels is an old one, with roots in both antiquity and the Middle Ages. As William Eamon demonstrates, however, it was used differently in different periods. In antiquity it connoted esoteric knowledge, which can only be discovered through supernatural means or revelation and should therefore be jealously guarded by the few initiates. In medieval scholasticism it referred to problems that were extremely difficult or simply unknowable. Such problems could not be solved through direct demonstration and were therefore excluded from the realm of proper "scientia." Finally, in the early modern period, the hidden gems and secrets of nature came to connote the unobservable "inner workings" of nature, which were to be exposed and brought to light. It is only this last sense that suggests that an exploration of nature is desirable. This is the sense in which Galileo, Cavalieri, and their colleagues are using the trope, while applying it to the study of mathematics. See Eamon, *Science and the Secrets of Nature* generally, the "Conclusion" for a summary, as well as his "Science as a Hunt."

38. On the new experimental sciences as a systematic search for hidden and undiscovered secrets, see Eamon, *Science and the Secrets of Nature*, chap. 8, as well as his "Science as a Hunt."

39. See also Alexander, "Imperialist Space of Elizabethan Mathematics."

40. Stephen J. Rigaud, *Correspondence of Scientific Men of the Seventeenth Century*, vol. 1 (Oxford: Oxford University Press, 1841), letter 24, 65–66.

41. On Baconianism in England, see Charles Webster, *The Great Instauration: Science, Medicine, and Reform, 1626–1660* (London: Duckworth, 1975); and Christopher Hill, *Intellectual Origins of the English Revolution* (Oxford: Clarendon Press, 1965).

42. Wallis published the entire correspondence related to his dispute with Fermat. See John Wallis, ed., *Commercium epistolicum de quaestionibus quibusdam mathematicis nuper habitum* (Oxford, 1658).

43. Not all mathematicians, however, accepted infinitesimals even then.

Descartes, after experimenting with them in his early years, abandoned them by the time he published the *Geometry* in 1638. The Jesuit order went even further and repeatedly banned the teaching of infinitesimals in its schools. On this, see Egidio Festa, "Quelques aspects de la controversie sur les indivisibles," in, *Geometria e atomismo nella scuola Galileana*, ed. Massimo Bucciantini and Maurizio Torrini (Florence: Olschki, 1992).

44. See Wallis's account of the work of Cavalieri, Hariot, and Oughtred in John Wallis, *A Treatise on Algebra, Both Historical and Practical* (London, 1685).

45. Peter Dear has recently emphasized the role of mathematics in the Royal Society's original program, described by John Wilkins as "Physico-Mathematicall-Experimentall learning." There is no doubt, however, that most of the work done under the auspices of the Royal Society in its early years was nonmathematical. See Dear, *Discipline and Experience*, esp. 2 and chap. 8.

46. See John Wallis, *Truth Tried or, Animadversions on a Treatise Published by the Right Honorable Robert Lord Brooke, Entitled the Nature of Truth*, printed by Richard Bishop for Samuel Gellibrand (London, 1642).

47. On his preference for sensible over abstract truths, see Wallis, *Truth Tried*, 60. The method of induction is used throughout his works, but is explicitly defended in his *Treatise of Algebra*, 306–8.

48. John Wallis, *Opera Mathematica*, vol. 1 (Oxford, 1699), 491.

49. On Bacon's pursuit, capture, and examination of nature through interrogation and torture, see Carolyn Merchant, *The Death of Nature: Women, Ecology, and the Scientific Revolution* (San Francisco: Harper & Row, 1980), chap. 7.

50. Wallis, *Opera Mathematica*, 1:491.

51. Bacon uses the hunt metaphor in his "De Sapientia Veterum," in *Works* 6:713 and in *De Augmentis Scientarum* 5.2, in *Works*, 1:633, trans. 4:421. See discussion in Eamon, *Secrets*, 283. On the image of judicial interrogation in Bacon, see discussions in Carolyn Merchant, *The Death of Nature* (San Francisco: Harper & Row, 1980), esp. chap. 7, and Evelyn Fox Keller, *Reflections on Gender and Science* (New Haven, Conn.: Yale University Press, 1985), chaps. 2 and 3.

52. These metaphors have often been treated interchangeably, both in the seventeenth century and in modern scholarship. Bacon uses all three tropes to argue his case for empiricism. Wallis, in the passage quoted, clearly uses the hunt and interrogation metaphors interchangeably. Paolo Rossi refers to the hunt and exploration tropes as interchangeable when he writes, "In this period there was continuous discussion with insistence that bordered on monotony about a logic of discovery conceived as a *venatio*, a hunt—as an attempt to penetrate territories never known or explored before." See Rossi, *Philosophy, Technology*, 42. Finally, William Eamon writes extensively about the trope of geographical exploration in his discussion of the hunt metaphor in early modern science. See Eamon, *Secrets*, chap. 8, 269–300.

A related but somewhat different theme was the notion of mathematics as an art for breaking secret codes. This notion gains added significance by the fact that many mathematicians, most notably Francois Viete and John Wallis, were indeed professional code breakers. There are, it seems to me, two versions of this trope: one, used by Viete and Descartes (among others), views mathematics as a tool for deciphering hidden secrets in the outside world. This version uses mathematics in new ways, but does not touch the core understanding of what mathematics is. It is used to refer, most often, to algebra. On this see Peter Pesic, "Secrets, Symbols, and Systems: Parallels between Cryptanalysis and Algebra, 1580–1700," *Isis* 88 (1997): 674–92. The other version, used by Edward Wright ("who for mathematics found the key . . .") and William Oughtred ("The Key to Mathematics"), implies that mathematics itself withholds secrets and that the role of the mathematician is to uncover them. This last version is very close to the "exploration" trope.

53. This, of course, was precisely the view of early modern reformers of knowledge on the role of the natural philosopher. On early modern science as an attempt to systematically uncover the hidden secrets of nature, see Eamon, *Science and the Secrets of Nature*, chap. 8, "Science as *Venatio*"; as well as his "Science as a Hunt."

54. Joseph Glanvill, *The Vanity of Dogmatizing* (London, 1661), 178.

55. The case is not much different in mixed mathematics than it is in pure mathematics. In astronomy, for example, one may describe the observed positions of the planets through the use of rigorous and deductive geometry. This in no way alters the fundamental nature of geometry as rigorous and deductive.

56. Galileo Galilei, *Le Opere di Galileo Galilei* (Florence: Edizione Nazionale, 1929–39), vol. 18, 67, letter no. 3889. Cavalieri was quoting from ode III of Horace's *Carmina*.

57. Torricelli, *Opere Scelte*, 382. From a modern perspective, the method of indivisibles may seem like a technique for solving problems rather than a "discovery" out there in the world. As the language here suggests, however, Cavalieri, Torricelli, Hariot, and their colleagues viewed the method very differently. For them it revealed something fundamental about the true nature of mathematical objects.

58. Cavalieri to Torricelli, March 10, 1643, letter no. 47 in *Opere de Torricelli*, 3:114.

59. Bonaventura Cavalieri, *Exercitationes geometricae sex* (Bologna, 1647).

60. While this conveys the general argument of the proof, certain details have been omitted in the interest of simplicity. Unlike Torricelli and other "indivisiblist" mathematicians, Cavalieri was a cautious practitioner of his method. He insisted, for example, that a single line ("regula") traverse both triangles simultaneously and insisted that the mathematical relation between the figures was valid only if it holds between every two lines that the regula intersects at the same moment. Furthermore, he never claimed that the lines actually comprise a plane figure (as did Torricelli) but only that if a certain relation holds between "all the lines" of one figure and "all the lines" of another, then the same relationship holds between

their surface areas. See François de Gandt, "Naissance et metamorphose d'une theorie mathematique: La géométrie des indivisibles en Italie," in *Sciences et Techniques en Perspective*, vol. 9 (Nantes: Universite de Nantes, 1984–85), 179–229; François de Gandt, "Les Indivisibles de Torricelli," in *L'Oeuvre de Torricelli: Science Galiléenne et nouvelle géométrie*, ed. F. de Gandt (Nice: Universite de Nice, 1987), 147–206; François de Gandt, "Cavalieri's Indivisibles and Euclid's Canons," in *Revolution and Continuity: Essays in the History and Philosophy of Early Modern Science*, ed. Peter Barker and Roger Ariew (Washington, D.C.: Catholic University of America Press, 1991), 157–82; and Kristi Andersen, "Cavalieri's Method of Indivisibles," *Archive for History of Exact Sciences* 31 (1985): 293–367.

61. Cavalieri was a cautious mathematician, much concerned with preserving the logical consistency of his system. He carefully avoided making any programmatic statements about the composition of the continuum, which would open him to criticism based on the ancient paradoxes of Zeno and the problem of incommensurability. Cavalieri, however, did have a clear idea of what the internal structure of geometrical figures was like. His basic intuition about the structure of the continuum is made clear in the opening passages of his *Exercitationes geometricae sex*, where he writes, "It is therefore evident that the plane figures should be conceived by us in the same manner as cloths are made up of parallel threads. And solids are in fact like books, which are composed of parallel pages" (*Exercitationes geometricae sex*, 3). Cavalieri's method of indivisibles was based on this fundamental intuition of the composition of the continuum.

62. Eamon, *Science and the Secrets of Nature*, 270.

63. See discussion in Enrico Giusti, *Bonaventura Cavalieri and the Theory of Indivisibles* (Cremona: Edizioni Cremonese, 1980). Guldin's critique of indivisibles is contained in his *Centrobaryca seu de centro gravitatis trium specierum quantitatis continuae*, 4 vols. (Vienna, 1635–41).

64. For the full discussion, see the *Two New Sciences*, 28–33.

65. Ibid., 35, 36, 38.

66. Ibid., 51.

67. Ibid., 54.

68. Ibid., 39.

69. On the respective approaches of Galileo, Cavalieri, and Torricelli to the Method of Indivisibles, see de Gandt, "Naissance et metamorphose d'une theorie mathematique."

70. For a fuller account of Torricelli's lists of paradoxes, see de Gandt, "Les indivisibles de Torricelli."

71. Ibid., 182.

72. I am here following de Gandt's argument on the nature of Torricelli's' mathematics in "Les indivisibles de Torricelli," esp. 181.

73. Quoted in L. Belloni, "Torricelli et son epoque (le triumvirat des eleves de Castelli: Maggiotti, Nardi, et Torricelli)" in *L'Ouevre de Torricelli: Science*

Galiléenne et nouvelle géométrie, publications de la faculte des lettres et sciences humaines de Nice no. 32 (first series), ed. François de Gandt (Nice: Les Belles Lettres, 1987), 31.

74. Nardi is referring to Archimedes' celebrated "method of exhaustion," which proved results on areas and volumes of geometrical objects deductively and without the use of infinitesimals. Essentially, this method proved that the area/volume of an object could not be greater or smaller than a given figure, because that would lead to a logical contradiction. Nardi (and many others) pointed out that while this method proved the result with certainty, it gave no hint as to how the result was originally reached. Many suspected that Archimedes had used infinitesimals as a method of discovery, before erasing all trace of them in his final proofs. Their suspicions did, of course, prove accurate with the discovery of Archimedes' "Method" by Heiberg at the turn of the twentieth century.

75. See, for example, Massimo Bucciantini and Maurizio Torrini, eds., *Geometria e Atomismo nella Scuola Galileana* (Florence: Olschki, 1992).

76. The biographical information on Stevin is from *Principal Works*, 1:3–14.

77. Stevin, *Principal Works*, 1:230. Also discussed in Carl B. Boyer, *The History of the Calculus and its Conceptual Development* (New York: Dover, 1959), 100.

78. Archimedes was the forerunner of infinitesimal techniques, including the "weighing" of geometrical figures and was acknowledged as such by sixteenth- and seventeenth-century mathematicians. In his formal presentations, however, Archimedes substituted these methods with rigorous and deductive proofs, the most famous of which was the "method of exhaustion." The early modern revivers of the infinitesimal methods found this substitution frustrating, as it did not reveal the original infinitesimalist investigations through which Archimedes discovered the results in the first place.

79. William Oughtred, preface to *The Key of Mathematics New Forged and Filed* (London, 1647). This English translation closely follows the Latin of the original *Clavis Mathematicae* published fifteen years earlier. See William Oughtred, *Clavis Mathematicae* (London, 1632).

80. Oughtred, preface to *The Key of Mathematics*.

81. Ibid.

82. Oughtred reiterates the same view of the purpose of mathematical studies in an introductory poem to his book entitled *Mathematicall Recreations* (London, 1653) where he writes:

> Here questions of ARITHMETRIK are wrought
> And hidden secrets unto light are brought

Here again mathematics withholds precious secrets, which are then brought to light by the mathematician.

83. On Oughtred, see J. F. Scott, "Oughtred, William," in *The Dictionary of Scientific Biography*, vol. 10 (New York: Scribner's, 1974); Florian Cajori,

William Oughtred, a Great Seventeenth Century Teacher of Mathematics (Chicago: Open Court, 1916); and Florian Cajori, "A List of Oughtred's Mathematical Symbols with Historical Notes," *University of California Publications in Mathematics* 1, no. 8 (February 18, 1920).

84. Rigaud, *Correspondence of Scientific Men*, 1:65–66. For a detailed discussion of Oughtred and his use of the language of exploration, see Chapter 3 above.

85. Ibid.

86. The visual imagery fits neatly with the exploration motif. Unlike the bookish scholars that remain in their studies, the explorer *views and observes* the wonders of unknown lands and then presents them to those who stayed behind.

87. Wallis, *Opera Mathematica*, 1:363.

88. It should be noted, though, that Wallis initially received his information on Cavalieri's method through Torricelli's *Opera Geometrica* (Florence, 1641). His conception of the method of indivisibles was accordingly much closer to Torricelli's notions than to Cavalieri's original method. See Wallis's introduction to his *Arithmetica Infinitorum* of 1656 in Wallis, *Opera Mathematica*, 1:357; and François de Gandt, "Les Indivisibles de Torricelli," 152.

89. Wallis's *Arithmetica Infinitorum* of 1656 was probably his most important mathematical work and immediately established his reputation as one of the leading mathematicians in Europe. See Wallis, *Opera Mathematica*, vol. 1.

90. Wallis, *Opera Mathematica*, 1:299.

91. On Fermat's reaction to Wallis, see his August 15, 1657, letter to Kenelm Digby, "Remarques sur l'Arithmetique des infinis du S. J. Wallis," in *Oeuvres de Fermat*, ed. Charles Henry and Paul Tannery (Paris: Gauthier-Villars, 1891–1912), 2:347–53; as well as Mahoney, *Fermat*, chap. 6.

92. Wallis, *Opera Mathematica*, 1:491. See n. 14 above.

93. Bernard de Fontenelle, *Elements de la géométrie de l'infini* (Paris, 1727), ciii; quoted in Michael S. Mahoney, "Infinitesimals and Transcendent Relations: The Mathematics of Motion in the Late Seventeenth Century," in *Reappraisals of the Scientific Revolution*, ed. David C. Lindberg and Robert S. Westman (Cambridge: Cambridge University Press, 1990), 461–91, n. 46. For a discussion of Fontenelle's views on infinity, see the epilogue of Michel Blay, *Reasoning with the Infinite: From the Closed World to the Mathematical Universe*, trans. M. B. DeBevoise (Chicago: University of Chicago Press, 1998). First published in French in 1993.

Conclusion

1. Georg Joachim Rheticus, "Narratio Prima," in *Three Copernican Treatises*, ed. Edward Rosen (New York: Dover, 1959).

2. Ibid., 191. The tale of Aristippus is narrated in Vitruvius's *De Architectura*, which was most likely Rheticus's source. See Vitruvius, *De Architectura*, preface to book VI. Aristippus was a sophist who lived in the fifth century BC.

3. John Davies of Hereford, "In praise of the never-too-much praised Worke and Author the L. of *Marchiston*," in John Napier, *A Description of the Admirable Table of Logarithms*, trans. Edward Wright (London, 1616). The "Lord of Marchiston" is John Napier, the inventor of logarithms. See discussion in Chapter 3 above.

4. For a discussion of science as the search for novelties never seen before, see William Eamon, *Science and the Secrets of Nature* (Princeton, N.J.: Princeton University Press, 1994), chap. 8; as well as Eamon, "Science as a Hunt," *Physis: Rivista Internazionale di Storia della Scienza* 32 (1994): 393–432. Eamon argues that modern science was born when the aims of science shifted from demonstrations of the familiar aspects of nature to a hunt for the unknown "secrets of nature." Eamon emphasizes the imagery of the hunt over that of geographical exploration, but his discussion and examples make it clear that the two images capture a similar vision of the purpose and practice of science.

5. This should be qualified in the case of "mixed mathematics," where "pure mathematics" was joined with natural philosophy. In those cases, of course, all truths could not be said to be derived from the initial assumptions. Nevertheless, as Clavius makes clear in the "Prolegomena," the basic nature of mathematics is not changed by combining it with natural philosophy: it maintains its strict logical and deductive character and bestows it on the new field. See discussion above.

6. It is probably no coincidence that just as French natural philosophers were more skeptical of empiricism than their English contemporaries, French mathematicians rarely employed the rhetoric of exploration and were far more cautious with the use of indivisibles. Viete did not discuss indivisibles at all, and Descartes famously excluded them from the realm of mathematics. Fermat used infinitesimals skillfully but gingerly, insisting that unlike Wallis he was proceeding in a "via ordinaria, legitimea, et Archimedea." On Fermat, see discussion in Michael S. Mahoney, *The Mathematical Career of Pierre de Fermat* (Princeton, N.J.: Princeton University Press, 1973), chap. 5. The quote is from a letter from Fermat to Kenelm Digby, published in Wallis's *Commercium Epistolicum* of 1658. See John Wallis, ed., *Commercium epistolicum de quaestionibus quibusdam mathematicis nuper habitum* (Oxford, 1658).

It is interesting to note that Fermat ultimately gave up his insistence on the classical nature of his practice in his later years, when it was clear that the new methods had far surpassed the classical model. It is in this context that we encounter the following passage in an introduction by Antoine de Lalouvere to Fermat's treatise on rectification, which he published as an appendix to his own book in 1660:

> hac tempestate in Geometricis inventum et superatum feliciter esse Bonae Spei promontorium illud, unde expedita existat navigatio ad inaccessas ante tetragonismorum praesertim regiones [At this time in geometry, the Cape of Good Hope was fortunately discovered and circumnavigated; whence navigation to regions practically inaccessible before the quadratures was made possible].

See Antoine de Lalouvere, *Veterum geometria promota* (Toulouse, 1660), part I, second appendix, preamble to some propositions of Fermat's. Lalouvere's text was reprinted in Pierre de Fermat, *Oeuvres de Fermat*, ed. Charles Henry and Paul Tannery (Paris: Gauthier-Villars, 1891–1912), 1:199–200, n. 1. I thank Michael Mahoney for this reference.

7. John Wallis, *A Treatise on Algebra, Both Historical and Practical* (London, 1685), 305.

8. Christopher Clavius, "Proemium de Algebrae Praestantia," in *Algebra Christophori Clavii Bambergensis* (Geneva, 1609), 1.

9. I do not mean to suggest that the perfectly self-consistent and coherent calculus sprang fully formed from the minds of Newton and Leibniz in the 1660s and 1680s. Clearly the calculus was still logically problematic and was criticized for its inconsistencies, most notably by Bishop Berkeley. Nevertheless, it did form a clear break from the haphazard infinitesimal methods that preceded it by emphasizing its internal self-consistent method, where its predecessors sought to deconstruct geometrical figures and the continuum into their infinitesimal components.

Appendix A

1. This is a paraphrase of J. G. A. Pocock's assessment of Christian eschatological history: "Secular Time . . . was the theater of redemption, but not its dimension." J. G. A. Pocock, *The Machiavellian Moment* (Princeton, N.J.: Princeton University Press, 1975), 8.

2. Karl Mannheim, *Ideology and Utopia* (New York: Harcourt, Brace & World, 1936), 71.

3. For an account of the development of the Sociology of mathematics, see part 2 of Sal Restivo, *The Social Relations of Physics, Mysticism, and Mathematics* (Boston: D. Reidel, 1983). On Marxist views, see the introduction to Richard W. Hadden, *On the Shoulders of Merchants* (New York: State University of New York Press, 1994).

4. Oswald Spengler, *The Decline of the West*, vol. 1 (New York: Alfred A. Knopf, 1950), 7. The work was first published in German as *Der Untergang des Abendlandes, Gestalt und Wirklichkeit* (Munich: Beck, 1918).

5. Spengler, *The Decline of the West*, 1:59; original emphasis. For a concise account of Spengler's views see chapter 1 of Sal Restivo, *Mathematics in Society and History* (Boston: Kluwer Academic Publishers, 1992).

6. Boris Hessen, "The Social and Economic Roots of Newton's *Principia*," in *Science at the Cross Roads* (London: Kniga, 1931).

7. My account of Borkenau is based on Richard W. Hadden, *On the Shoulders of Merchants*. See also Franz Borkenau, *Der Übergang vom feudalen zum bürgerlichen Weltbild* (1934; reprint, Darmstadt: Wissenschaftliche Buchgesellschaft, 1976).

8. Hadden, *On the Shoulders of Merchants*.

9. For Bloor's account of Wittgenstein, see David Bloor, "Wittgenstein and Mannheim on the Sociology of Mathematics," *Studies in the History and Philosophy of Science* 4 (1973): 173–91. For his discussion of John Stuart Mill and his views on mathematics, see David Bloor, *Knowledge and Social Imagery* (Chicago: University of Chicago Press, 1991). For Bloor's sources, see Ludwig Wittgenstein, *Remarks on the Foundations of Mathematics* (Cambridge, Mass.: Harvard University Press, 1967); and John Stuart Mill, *A System of Logic: Ratiocinative and Inductive* (New York: Harper, 1846).

10. Restivo, *Mathematics in Society*. Restivo deals with a large variety of approaches and different issues in his book, but I take this thesis to be a major theme that runs through the book.

11. Oswald Spengler's mystical approach to culture and numbers is an exception. I find some of his insights into the cultural nature of mathematics to be intriguing and worthy of further investigation. Spengler, however, has been intellectually delegitimized in the post-war period due to the unsavory nature of his political affiliations. His work on mathematics still awaits serious scholarly consideration.

12. Bloor, *Knowledge and Social Imagery*, 89–92.

13. Philip Kitcher, *The Nature of Mathematical Knowledge* (Oxford: Oxford University Press, 1983).

14. The discussion here is not meant to provide a comprehensive survey of the historiography of mathematics. The authors discussed offer but a sample of the different approaches to historicizing mathematics that have been attempted over the years. They are meant to represent leading trends, but not to cover all of them.

Other interesting approaches to the questions have been suggested in recent years. Timothy Lenoir has suggested viewing developments in late-nineteenth- and early-twentieth-century mathematics against the philosophical background of the emergence of German phenomenology. Joan L. Richards interprets the spread of analysis and non-Euclidean geometry in Victorian England in terms of the pedagogical reform of English universities and public schools and the philosophical convictions of the leading reformers. Brian Rotman proposes a semiotic interpretation of mathematics, in which mathematical practices conform to contemporary semiotic guidelines prevailing in other fields. Paul Ernest has advocated a reform in the teaching of mathematics based on a social constructivist understanding of mathematics. Reuben Hersh advocates a "humanistic" understanding of mathematics, as part of human culture and history. See Timothy Lenoir, "Practical Reason and the Construction of Knowledge: The Life-World of Haber Bosch," in *The Social Dimensions of Science*, ed. Ernan McMullin (South Bend, Ill.: University of Notre Dame Press, 1992), 158–97; Joan L. Richards, *Mathematical Visions: The Pursuit of Geometry in Victorian England* (San Diego: Academic Press, 1988); Brian Rotman, *Signifying Nothing: The Semiotics of Zero* (Stanford, Calif.: Stanford University Press, 1993); Brian Rotman, *Ad Infinitum: The Ghost in Turing's Machine* (Stanford, Calif.: Stanford University Press, 1993); Paul Ernest, *Social*

Constructivism as a Philosophy of Mathematics (Albany: State University of New York Press, 1998); and Reuben Hersh, *What Is Mathematics Really?* (Oxford: Oxford University Press, 1997).

15. My account of the formalist approach to mathematics is based on Stephen C. Kleene, *Introduction to Metamathematics* (Amsterdam: North Holland Publishing, 1964), 24–29, 53–65; and Mary Tiles's excellent account of Hilbert's program in Mary Tiles, *Mathematics and the Image of Reason* (London: Routledge, 1991). Hilbert gave an early example of his approach in his *Grundlagen der Géométrie* of 1899. For an English version, see David Hilbert, *Foundations of Geometry*, trans. Leo Unger (La Salle, Ill.: Open Court, 1971).

16. For an account of non-Euclidean geometry and its profound impact on the traditional views on mathematics, see Richards, *Mathematical Visions*.

17. Restivo, *Mathematics in Society*, 167.

18. See, for example, the development of the calculus as a nonrigorous and paradoxical technique in the seventeenth and eighteenth centuries. This episode in the history of mathematics is recounted at the beginning of Chapter 5 and in Chapter 6 above.

19. My use of the concept of narrative is similar to James Bono's views of the function of metaphor in science. According to Bono, metaphors are constitutive of scientific research and knowledge, but they are also culturally grounded. Science is therefore culturally and historically situated through its use of metaphors. The challenge of the field of literature and science, according to Bono, is to recapture the connections between scientific and nonscientific discourses by tracing their use of metaphors. My own "narrative," much like Bono's "metaphor," is a culturally specific literary device, which can therefore be used to establish the fundamental historicity of a mathematical/scientific practice. See James J. Bono, "Science, Discourse, and Literature: The Role/Rule of Literature in Science," in *Literature and Science: Theory and Practice*, ed. Stuart Peterfreund (Boston: Northeastern University Press, 1990), 59–90.

20. A few recent works applying similar approaches to other branches of science should be mentioned: Donna Haraway's *Primate Visions* demonstrates how cultural narratives on topics such as decolonization and the cold war became encoded into the study of primatology. Gillian Beer traced the literary plots used by Darwin in his *Origin of the Species* and relates them to nineteenth-century literary trends. See Gillian Beer, *Darwin's Plots: Evolutionary Narratives in Darwin, George Eliot, and Nineteenth Century Fiction* (London: Routledge, 1983). James Bono provides an excellent methodological framework for relating scientific and literary works in his "Science, Discourse, and Literature."

21. There are, of course, obvious gendered overtones to this narrative, which I do not pursue here. For the sexual significance of division into parts in the English Renaissance, see Patricia Parker, "Rhetorics of Property: Exploration, Inventory, Blazon," in her *Literary Fat Ladies* (London: Methuen, 1987); and Nancy Vickers, "Diana Described," *Critical Inquiry* 8 (1981): 265–79. On the gendered content

of exploration narratives, see Louis Montrose, "The Work of Gender in the Discourse of Discovery," *Representations* 33 (winter 1991): 1–41; and Annette Kolodny, *The Lay of the Land* (Chapel Hill: University of North Carolina Press, 1975). For general discussions of gender and science in this period, see Evelyn Fox Keller, *Reflections on Gender and Science* (New Haven, Conn.: Yale University Press, 1985); and Carolyn Merchant, *The Death of Nature* (San Francisco: Harper & Row, 1980).

Appendix B

1. For a more detailed and technical account of Hariot's work on the Mercator projection, see Jon V. Pepper, "Harriot's Calculation of the Meridional Parts as Logarithmic Tangents," *Archive for the History of Exact Sciences* 4 (1967–68): 359–413.

2. He found that the value of β, in modern decimal notation, was

$\beta = 0.9970915409725778.$

3. Hariot, in fact, did not settle for the β minutes table, but produced separate tables for whole degrees and 10ths of minutes as well. When comparing his tables with the values of $\tan(\Phi/4 - \phi/2)r_0$ he started out with the "degree table," moved to the "minutes table" and eventually to the "10th of a minute" table, thus determining the meridional parts more efficiently. See Pepper, "Harriot's Calculation." Pepper also describes the sophisticated interpolation techniques used by Hariot in constructing the table for 1/10th of a minute.

BIBLIOGRAPHY

Manuscript Sources

The Hariot Papers in the British Library, B.L. Add. MSS 6782–89.
The Hariot Papers in Petworth House, Sussex, HMC 240/i–v, HMC 241/i–x.
Map of Guyana attributed to Thomas Hariot, B.L. Add. MS 17940.
"A navigational rutter with notes by Thomas Hariot," B.L. Sloane MS 2292.
Hariot, Thomas, "De Reflexione," Petworth House MS. HMC 241/iv, ff. 23–31, and B.L. Harley MS 6002, ff. 16v–20.
Dee, John, "Britanici Imperii Limites," B.L. Add. MS 59681.
Torporley, Nathaniel, "Synopsis of the controversie of Atoms," B.L. Birch MS 4458 ff. 6–8.

Printed Sources

Alexander, Amir R. "The Imperialist Space of Elizabethan Mathematics." *Studies in the History and Philosophy of Science* 26 (1995): 559–91.
———. "Lunar Maps and Coastal Outlines: Thomas Hariot's Mapping of the Moon." *Studies in the History and Philosophy of Science* 29 (1998): 345–68.
Andersen, Kristi. "Cavalieri's Method of Indivisibles." *Archive for History of Exact Sciences* 31 (1985): 293–367.
Andrews, K. R. *Trade, Plunder, and Settlement.* Cambridge: Cambridge University Press, 1984.
Aristotle. *De Generatione et Corruptione.* Ed. C. J. F. Williams. Oxford: Clarendon Press, 1982.
———. *The Works of Aristotle Translated into English.* 12 vols. Ed. W. D. Ross. Oxford: Clarendon Press, 1930.
Armesto, Felipe Fernandez. *Columbus.* Oxford: Oxford University Press, 1991.
Aspley, John. *Speculum Nauticum: A Looking Glass for Sea-Men.* London, 1624.
Bacon, Francis. *The New Organon and Related Writings.* New York: Liberal Arts Press, 1960.

————. *The Works of Francis Bacon.* 14 vols. Ed. James Spedding, Robert Leslie Ellis, and Douglas Denon Heath. London: Longman, 1858–74.

Bacon, Roger. "Opus Tertium." In *Rerum Britannicorum Medii Aevi Scriptores (The Rolls Series)*, vol. 15, ed. J. S. Brewer, chap. 39, 131–35. London: H. M. Stationery Office, 1859.

Baron, Margaret E. *The Origins of the Infinitesimal Calculus.* New York: Pergamon Press, 1969.

Barthes, Roland. *Mythologies.* Ed. and trans. Annette Lavers. New York: Noonday Press, 1972.

Beer, Gillian. *Darwin's Plots: Evolutionary Narratives in Darwin, George Eliot, and Nineteenth Century Fiction.* London: Routledge, 1983.

Belloni, L. "Torricelli et son epoque (le triumvirat des eleves de Castelli: Maggiotti, Nardi, et Torricelli)." In *L'Ouevre de Torricelli: Science Galiléenne et nouvelle géométrie*, ed. F. de Gandt, 29–38. Nice: Publications de la faculte des lettres et sciences humaines de Nice no. 32 (first series), 1987.

Blay, Michel. *Reasoning with the Infinite: From the Closed World to the Mathematical Universe.* Trans. M. B. DeBevoise. Chicago: University of Chicago Press, 1998. First published in French in 1993.

Bloor, David. *Knowledge and Social Imagery.* Chicago: University of Chicago Press, 1991.

————. "Wittgenstein and Mannheim on the Sociology of Mathematics." *Studies in the History and Philosophy of Science* 4 (1973): 173–91.

Blundeville, Thomas. *His Exercises, Containing Eight Treatises.* London, 1606.

————. *The Theoriques of the Seven Planets.* London, 1602.

————. *The Theoriques of the Seven Planets, shewing all their diverse motions, and all other Accidents, called Passions, thereunto belonging.* London, 1602.

Bollaert, William. *The Expedition of Pedro de Ursua & Lope de Aguirre in Search of El Dorado and Omagua in 1560–1.* London: Hakluyt Society Publications 28, 1861.

Bono, James J. "Science, Discourse, and Literature: The Role/Rule of Literature in Science." In *Literature and Science: Theory & Practice*, ed. Stuart Peterfreund, 59–90. Boston: Northeastern University Press, 1990.

Borkenau, Franz. *Der Übergang vom feudalen zum bürgerlichen Weltbild.* 1934. Reprint, Darmstadt: Wissenschaftliche Buchgesellschaft, 1976.

Bostock, David. "Aristotle on Continuity in *Physics* VI." In *Aristotle's Physics: A Collection of Essays*, ed. Lindsay Judson, 179–212. Oxford: Clarendon Press, 1991.

Boyer, Carl B. *The History of the Calculus and its Conceptual Development.* New York: Dover Publications, 1959.

————. *The Rainbow, from Myth to Mathematics.* Princeton, N.J.: Princeton University Press, 1987.

Boyer, Carl B., and Uta C. Merzbach. *A History of Mathematics.* 2d ed. New York: John Wiley & Sons, 1991.

Briggs, Henry. "A Treatise of the Northwest Passage to the South Sea, Through the Continent of Virginia and by Fretum Hudson." In *A Declaration of the State of the Colony in Virginia*, ed. Edward Waterhouse. London, 1622.

Bruno, Giordano. *Ash Wednesday Supper*. Trans. Stanley L. Jaki. The Hague: Mouton, 1975. First published as *Cena de le Ceneri*, London, 1584.

Bry, Theodor de. *Discovering the New World, Based on the Work of Theodore de Bry*. Ed. Michael Alexander. New York: Harper & Row, 1976.

Bucciantini, Massimo, and Maurizio Torrini, eds. *Geometria e Atomismo nella Scuola Galileana*. Florence: Olschki, 1992.

Buisseret, David, ed. *Monarchs, Ministers, and Maps: The Emergence of Cartography as a Tool of Government in Early Modern Europe*. Chicago: University of Chicago Press, 1992.

Cajori, Florian. "On an Integration ante-dating the Integral Calculus." *Bibliotheca Mathematica* 14 (1914–15): 312–19.

———. "A List of Oughtred's Mathematical Symbols, with Historical Notes." *University of California Publications in Mathematics* 1, no. 8 (February 18, 1920).

———. "Oughtred's Ideas and Influence on the Teaching of Mathematics." *The Monist* 25 (1915): 495–530.

———. *William Oughtred, A Great Seventeenth-Century Teacher of Mathematics*. Chicago: Open Court, 1916.

Campbell, Mary B. *The Witness and the Other World*. Ithaca, N.Y.: Cornell University Press, 1988.

Cavalieri, Bonaventura. *Exercitationes geometricae sex*. Bologna, 1647.

———. *Geometria indivisibilibus continuorum nova quadam ratione promota*. Bologna, 1635.

Cawley, Robert Ralston. *Unpathed Waters*. Princeton, N.J.: Princeton University Press, 1940.

Chapman, George. *The Poems of George Chapman*. Ed. Phyllis B. Bartlett. Oxford: Oxford University Press, 1941.

Christy, Miller, ed. *The Voyages of Captain Luke Foxe of Hull and Captain Thomas James of Bristol*. London: Hakluyt Society Publications 88–89, 1894.

Clavius, Christopher. "In disciplinas mathematicas prolegomena." In *Opera Mathematica*, by Christopher Clavius, vol. 1. Mainz, 1611.

———. "Proemium de algebrae praestantia." In *Algebra Christophori Clavii Bambergensis*, by Christopher Clavius, 1–3. Geneva, 1609.

Clulee, N. H. *John Dee's Natural Philosophy*. London and New York: Routledge, 1988.

Collinson, Richard. *Three Voyages of Martin Frobisher*. London: Hakluyt Society Publications 38, 1867.

Columbus, Christopher. *The Diario of Christopher Columbus's First Voyage to America, Abstracted by Fray Bartolomé de Las Casas*. Ed. Oliver Dunn and James E. Kelley Jr. Norman: University of Oklahoma Press, 1988.

———. *The Libro de las profecias of Christopher Columbus*. Trans. and ed. Delano C. West and August Kling. Gainsville: University of Florida Press, 1991.

———. "The Will of Christopher Columbus." In *The Authentic Letters of Columbus*, ed. William Eleroy Curtis, 193–200. Chicago: Field Columbian Museum, 1895.

Cormack, Lesley. "Britannia Rules the Waves? Images of Empire in Elizabethan England." *Early Modern Literary Studies* (electronic journal), September 1998, (http://www.humanities.ualberta.ca/emls/04-2/cormbrit.html).

Cortés, Hernán. *Letters from Mexico*. Trans. and ed. A. Pagden. New Haven, Conn.: Yale University Press, 1986.

Cumming, P., R. A. Skelton, and D. B. Quinn. *The Discovery of North America*. New York: American Heritage Press, 1972.

Davis, John. "The Seamans Secrets." In *The Voyages and Works of John Davis the Navigator*, ed. Albert Hastings Markham, 224–337. London: Hakluyt Society Publications 59, 1880.

———. *The Worldes Hydrographical Description*. London, 1595.

Dear, Peter. *Discipline and Experience: The Mathematical Way in the Scientific Revolution*. Chicago: University of Chicago Press, 1995.

Dee, John. *General and Rare Memorials Pertayning to the Perfect Arte of Navigation*. London, 1577.

Dijksterhuis, E. J. *Archimedes*. 1938. Reprint, Princeton, N.J.: Princeton University Press, 1987.

———. *Simon Stevin, Science in the Netherlands Around 1600*. The Hague: Nijhoff, 1970.

Eamon, William. "Science as a Hunt." *Physis: Rivista Internazionale di Storia della Scienza* 31 (new series) (1994): 393–432.

———. *Science and the Secrets of Nature*. Princeton, N.J.: Princeton University Press, 1994.

Easton, Joy B. "Recorde, Robert." In *The Dictionary of Scientific Biography*, 16 vols., ed. Charles C. Gillispie. New York: Scribners, 1970–80.

———. "A Tudor Euclid." *Scripta Mathematica* 27 (1966): 339–55.

Edwards, Edward. *The Life of Sir Walter Raleigh Together with his Letters*. 2 vols. London: Macmillan & Co., 1868.

Elliott, J. H. *The Old World and the New 1492–1650*. Cambridge: Cambridge University Press, 1970.

———. "Renaissance Europe and America: A Blunted Impact?" In *First Images of America*, vol. 1, ed. Fredi Chiappelli, 11–23. Berkeley and Los Angeles: University of California Press, 1976.

Ernest, Paul. *Social Constructivism as a Philosophy of Mathematics*. Albany: State University of New York Press, 1998.

Feingold, Mordechai. *The Mathematicians' Apprenticeship: Science, Universities and Society in England, 1560–1640*. Cambridge: Cambridge University Press, 1984.

Ferguson, Margaret W., Maureen Quilligan, and Nancy J. Vickers, eds. *Rewriting the Renaissance*. Chicago: University of Chicago Press, 1986.

Fermat, Pierre de. *Oeuvres de Fermat*. 4 vols. Ed. Charles Henry and Paul Tannery. Paris: Gauthier-Villars, 1891–1922.

Festa, Egidio. "Quelques aspects de la controversie sur les indivisibles." In *Geometria e atomismo nella scuola Galileana*, ed. Massimo Bucciantini and Maurizio Torrini, 193–207. Florence: Olschki, 1992.

Findlen, Paula. "Il nuovo Colombo: Conoscenza e ignoto nell'Europa del Rinascimento." In *La rappresentazione dell'altro nei testi del Rinascimento*, ed. Sergio Zatti, 219–44. Lucca: M. Pazzini Fazzi, 1998.

Fontenelle, Bernard de. *Elements de la Géométrie de l'infini*. Paris, 1727.

Franklin, Wayne. *Discoverers, Explorers, Settlers*. Chicago: University of Chicago Press, 1979.

French, Peter J. *John Dee: The World of an Elizabethan Magus*. London: Routledge, 1972.

Fuller, Mary C. "Raleigh's Fugitive Gold: Reference and Deferral in *The Discoverie of Guiana*." *Representations* 33 (1991): 42–64.

Galilei, Galileo. *Dialogue Concerning the Two Chief World Systems*. Trans. and ed. Stillman Drake. Berkeley and Los Angeles: University of California Press, 1967.

———. *Le Opere di Galileo Galilei*. 20 vols. Florence: G. Barbèra, 1929–39.

———. *Two New Sciences*. Trans. and ed. Stillman Drake. Madison: University of Wisconsin Press, 1974.

Galinsky, Hans. "Exploring the 'Exploration Report' and its Image of the Overseas World: Spanish, French, and English Variants of a Common Form Type in Early American Literature." *Early American Literature* 12 (1977): 5–24.

Gandt, François de. "Cavalieri's Indivisibles and Euclid's Canons." In *Revolution and Continuity: Essays in the History and Philosophy of Early Modern Science*, ed. Peter Barker and Roger Ariew, 157–82. Washington, D.C.: The Catholic University of America Press, 1991.

———. "Les Indivisibles de Torricelli." In *L'Oeuvre de Torricelli: Science Galiléenne et nouvelle géométrie*, ed. F. de Gandt, 147–206. Nice: Universite de Nice, 1987.

———. "Naissance et metamorphose d'une theorie mathematique: la geometrie des indivisibles en Italie." In *Sciences et Techniques en Perspective*, vol. 9, 179–229. Nantes: Universite de Nantes, 1984–85.

Gatti, Hilary. *The Natural Philosophy of Thomas Harriot*. Oxford: Oriel College, 1993.

———. *The Renaissance Drama of Knowledge*. New York: Routledge, 1989.

Gilbert, Humphrey. *A Discourse of the Discoverie for a New Passage to Cathaia*. London, 1576.

Gilbert, William. *De Magnete*. London, 1600.

Gillot, Herbert. *La querelle des anciens et des modernes en France*. Paris: Librairie Ancienne H. Champion & É. Champion, 1914.

Giusti, Enrico. *Bonaventura Cavalieri and the Theory of Indivisibles*. Cremona: Edizioni Cremonese, 1980.

Glanvill, Joseph. *The Vanity of Dogmatizing*. London, 1661.

Goodman, Jennifer R. *Chivalry and Exploration*. Woodbridge: Boydell Press, 1998.

Grafton, Anthony. *New Worlds Ancient Text: The Power of Tradition and the Shock of Discovery*. Cambridge, Mass.: The Belknap Press of Harvard University Press, 1992.

Greenblatt, Stephen. *Marvelous Possessions*. Chicago: University of Chicago Press, 1991.

Guldin, Paul. *Centrobaryca seu de centro gravitatis trium specierum quantitatis continuae*. 4 vols. Vienna, 1635–41.

Hadden, Richard W. *On the Shoulders of Merchants*. Albany: State University of New York Press, 1994.

Hakewill, George. *An Apologie or Declaration of the Power and Providence of God in the Government of the World*. 2d ed. Oxford, 1630.

Hakluyt, Richard. *The Principal Navigations Voyages, Traffiques and Discoveries of the English Nation, in Twelve Volumes*. Glasgow: MacLehose, 1903–5.

Haraway, Donna. *Primate Visions*. New York: Routledge, 1989.

Harcourt, Robert. *A Relation of a Voyage to Guiana*. London: Hakluyt Society Publications 2, no. 60, 1928.

Hariot, Thomas. *Artis Analyticae Praxis*. London, 1631.

———. *A Briefe and True Report of the New Found Land of Virginia*. New York: Dover Publications, 1972.

Harley, J. B. "Silences and Secrecy: The Hidden Agenda of Cartography in Early Modern Europe." *Imago Mundi* 40 (1989): 58–76.

Harlow, V. T., ed. *Raleigh's Last Voyage*. London: Argonaut Press, 1932.

Harvey, Richard. *A Theological Discourse of the Lamb of God*. London, 1590.

Helgerson, Richard. *Forms of Nationhood: The Elizabethan Writing of England*. Chicago: University of Chicago Press, 1992.

———. "The Land Speaks: Cartography, Chorography, and Subversion in Renaissance England." *Representations* 16 (fall 1986): 51–85.

Henry, John. "Thomas Harriot and Atomism: A Reappraisal." *History of Science* 20 (1982): 267–96.

Hersh, Reuben. *What Is Mathematics Really?* Oxford: Oxford University Press, 1997.

Hessen, Boris. "The Social and Economic Roots of Newton's *Principia*." In *Science at the Cross Roads*, by Boris Hessen. London: Kniga, 1931.

Hilbert, David. *Foundations of Geometry*. Trans. Leo Unger. La Salle, Ill.: Open Court, 1971.

Hill, Christopher. *Intellectual Origins of the English Revolution*. Oxford: Clarendon Press, 1965.

Holland, Cliff. *Arctic Exploration and Development: An Encyclopedia.* New York: Garland Publishing, 1994.

Housley, Norman. *The Later Crusades.* Oxford: Oxford University Press, 1992.

Hues, Robert. *Tractatus de Globis.* Ed. Clement R. Markham. London, 1592. Reprint, London: Hakluyt Society Publications, no. 79, 1889.

Hulton, Paul, and David B. Quinn. *The American Drawings of John White.* 2 vols. London: The British Museum, 1964.

Jacquot, Jean. "Humanisme et science dans l'Angleterre Elisabethaine: L'oeuvre de Thomas Blundeville." *Revue d'Histoire des Sciences* 6 (1953): 189–202.

———. "Thomas Hariot's Reputation for Impiety." *Notes and Records of the Royal Society of London* 9 (1952): 164–87.

Jane, Cecil, ed. *The Four Voyages of Columbus.* New York: Dover, 1988.

Jessopp, Augustus. "Blundeville, Thomas." In *Dictionary of National Biography,* vol. 5, ed. Leslie Stephen and Sidney Lee. London: Smith, Elder & Co., 1886.

Johnson, Francis R., and Sanford V. Larkey. "Robert Recorde's Mathematical Teaching and the Anti Aristotelian Movement." *Huntington Library Bulletin* 7 (April 1935).

Kalmar, Martin. "Thomas Harriot's 'De Reflexione Corporum Rotundorum': An Early Solution to the Problem of Impact." *Archive for the History of Exact Sciences* 16 (1977): 201–30.

Kargon, Robert. *Atomism in England from Harriot to Newton.* Oxford: Clarendon Press, 1966.

Keller, Evelyn Fox. *Reflections on Gender and Science.* New Haven, Conn.: Yale University Press, 1985.

Kenyon, W. A. *Tokens of Possession: The Northern Voyages of Martin Frobisher.* Toronto: Royal Ontario Museum, 1975.

Kepler, Johannes. *Ad Vitellionem paralipomena.* Frankfurt am Main, 1604.

———. *Johannes Keplers Gesammelte Werke.* 20 vols. Munich: C. H. Beck, 1937–98.

———. *Nova stereometria doliorum vinariorum.* Linz, 1615.

Keymis, Lawrence. *A Relation of the Second Voyage to Guiana.* London, 1596. Reprint, Amsterdam: Theatrum Orbis Terrarum, 1968.

Kitcher, Philip. *The Nature of Mathematical Knowledge.* Oxford: Oxford University Press, 1983.

Kleene, Stephen C. *Introduction to Metamathematics.* Amsterdam: North Holland Publishing, 1964.

Kline, Morris. *Mathematical Thought from Ancient to Modern Times.* Oxford: Oxford University Press, 1972.

Knorr, Wilbur R. "Infinity and Continuity: The Interaction of Mathematics and Philosophy in Antiquity." In *Infinity and Continuity in Ancient and Medieval Thought,* ed. Norman Kretzmann, 87–111. Ithaca, N.Y.: Cornell University Press, 1982.

Kolodny, Annette. *The Lay of the Land*. Chapel Hill: University of North Carolina Press, 1975.

Lalouvere, Antoine de. *Veterum geometria promota*. Toulouse, 1660.

Las Casas, Bartolome de. *In Defense of the Indians*. Trans. Stafford Poole. De Kalb: Northern Illinois University Press, 1974.

Lattis, James A. *Between Copernicus and Galileo: Christoph Clavius and the Collapse of Ptolemaic Astronomy*. Chicago: University of Chicago Press, 1994.

Lejeune, Albert. *Recherches sur la catoptrique grecque*. Brussels: Académie royale des sciences, 1957.

Lenoir, Timothy. "Practical Reason and the Construction of Knowledge: The Life-World of Haber Bosch." In *The Social Dimensions of Science*, ed. Ernan McMullin, 158–97. South Bend, Ill.: University of Notre Dame Press, 1992.

———. *The Social and Intellectual Roots of Discovery in Seventeenth Century Mathematics*. Unpublished doctoral dissertation, Indiana University, 1974.

Letts, Malcolm, ed. *Mandeville's Travels*. London: The Hakluyt Society Publications, series 2, nos. 101–2, 1953.

Lindberg, David C. "The Cause of Refraction in Medieval Optics." *British Journal for the History of Science* 4 (1968–69): 23–38.

———. "The Genesis of Kepler's Theory of Light: Light Metaphysics from Plotinus to Kepler." *Osiris*, 2d series, vol. 2 (1986): 5–42.

———. *Theories of Vision from Al-Kindi to Kepler*. Chicago: University of Chicago Press, 1976.

Lindberg, David C., and Robert S. Westman, eds. *Reappraisals of the Scientific Revolution*. Cambridge: Cambridge University Press, 1990.

Lohne, Johannes A. "Essays on Thomas Harriot: I. Billiard Balls and the Laws of Collision." *Archive for the History of Exact Sciences* 20 (1979): 189–204.

———. "The Fair Fame of Thomas Harriott: Rigaud versus Baron von Zach." *Centaurus* 8 (1963): 69–84.

———. "Harriot, Thomas." In *The Dictionary of Scientific Biography*. 16 vols., ed. Charles C. Gillispie. New York: Scribners, 1970–80.

———. "Kepler und Harriot, Ihre Wege zum Berchungsgesetz." In *Internationales Kepler-Symposium, Weil der Stadt, 1971*, ed. Fritz Krafft, Karl Meyer, and Bernhard Sticker. Hildesheim: Gerstenberg, 1973.

———. "Thomas Harriott (1560–1621): The Tycho Brahe of Optics." *Centaurus* 6 (1959): 113–21.

Mahoney, Michael S. *The Mathematical Career of Pierre de Fermat*. Princeton, N.J.: Princeton University Press, 1973.

Mancosu, Paolo. "Aristotelian Logic and Euclidean Mathematics: Seventeenth Century Developments of the Quaestio de Certitudine Mathematicarum." *Studies in the History and Philosophy of Science* 23 (1992): 241–65.

———. *Philosophy of Mathematics and Mathematical Practice in the Seventeenth Century*. Oxford: Oxford University Press, 1996.

Mannheim, Karl. *Ideology and Utopia*. New York: Harcourt, Brace & World, 1936.

Markham, Albert H. *The Voyages and Works of John Davis the Navigator.* London: Hakluyt Society Publications 59, 1880.

Merchant, Carolyn. *The Death of Nature: Women, Ecology, and the Scientific Revolution.* San Francisco: Harper & Row, 1980.

Mill, John Stuart. *A System of Logic: Ratiocinative and Inductive.* New York: Harper, 1846.

Mitchell, W. J. T., ed. *Landscape and Power.* Chicago: University of Chicago Press, 1994.

Montrose, Louis. "The Work of Gender in the Discourse of Discovery." *Representations* 33 (winter 1991): 1–41.

Morison, Samuel Eliot. *The European Discovery of America.* 2 vols. New York: Oxford University Press, 1971.

Murdoch, John E. "Infinity and Continuity." In *The Cambridge History of Later Medieval Philosophy*, ed. Norman Kretzmann, Anthony Kenny, and Jan Pinborg, 564–91. Cambridge: Cambridge University Press, 1982.

Murdoch, John E. "Naissance et developement de l'atomisme au bas moyen age Latin." In *La science de la nature: theories et pratique*, Cahiers d'etudes medievales 2. Montreal: Bellarmin, 1974.

——. "William of Ockham and the Logic of Infinity and Continuity." In *Infinity and Continuity in Ancient and Medieval Thought*, ed. Norman Kretzmann, 165–206. Ithaca, N.Y.: Cornell University Press, 1982.

Napier, John. *A Description of the Admirable Table of Logarithms.* Trans. Edward Wright. London, 1616.

——. *Mirifici logarithmorum canonis descriptio eiusque usus, ut etiam in trigonometria, ut etiam in omni logistica mathematica amplissimi, facillimi, & expedissimi explicatio.* Edinburgh, 1614.

Nashe, Thomas. *Pierce Penilesse, his Supplication to the Devill.* London, 1592.

Nicholl, Charles. *The Creature in the Map.* New York: William Morrow & Company, 1996.

North, J. D. "Finite and Otherwise: Aristotle and Some Seventeenth Century Views." In *Nature Mathematized*, ed. William R. Shea, 113–48. Dordrecht: D. Reidel, 1983.

——. "Thomas Harriot's Papers on the Calendar." In *The Light of Nature: Essays in the History and Philosophy of Science Presented to A. C. Crombie*, ed. John D. North and John J. Roche, 145–74. Dordrecht: M. Nijhoff, 1985.

Oughtred, William. *The Circles of Proportion and the Horizontall Instrument.* London, 1632.

——. *Clavis Mathematicae.* London, 1632.

——. *Mathematicall Recreations.* London, 1653.

——. *The Key of Mathematicks New Forged and Filed.* London, 1647.

Oviedo Y Valdez, Gonzalo Fernandez de. *Historia general y natural de las Indias.* Ed. Buesa Juan Perez di Tudela. Madrid: Biblioteca de autores espanoles, vols. 117–21, 1959.

———. *Libro de muy esforcado y invencible Cavallero della fortuna propriamente llamado don claribalte*. Valencia, 1519.

Pacius, Julius. *Aristotelis naturalis ausculationis*. Frankfurt am Main, 1596.

Pagden, Anthony. *European Encounters with the New World*. New Haven, Conn.: Yale University Press, 1992.

———. "Ius et Factum: Text and Experience in the Writings of Bartolomé de Las Casas." In *New World Encounters*, ed. Stephen Greenblatt, 85–100. Berkeley and Los Angeles: University of California Press, 1993. First published in *Representations* 33 (winter 1991): 147–62.

———. *Lords of All the World: Ideologies of Empire in Spain, Britain, and France, c. 1500–1800*. New Haven, Conn.: Yale University Press, 1995.

Parker, Patricia. *Literary Fat Ladies*. London: Methuen, 1987.

Parry, J. H. *The Age of Reconnaissance*. Berkeley and Los Angeles: University of California Press, 1981.

Pepper, Jon V. "Harriot's Calculation of the Meridional Parts as Logarithmic Tangents." *Archive for the History of Exact Sciences* 4 (1967–68): 359–413.

———. "Harriot's Earlier Work on Mathematical Navigation: Theory and Practice." In *Thomas Harriot, Renaissance Scientist*, ed. John W. Shirley, 54–90. Oxford: Clarendon Press, 1974.

———. "Harriot's Manuscript on the Theory of Impacts." *Annals of Science* 33 (1976): 131–51.

———. "Harriot's Unpublished Papers." *History of Science* 6 (1968): 17–40.

———. "Some Clarifications of Harriot's Solution of Mercator's Problem." *History of Science* 14 (1976): 235–44.

Pesic, Peter. "Secrets, Symbols, and Systems: Parallels between Cryptanalysis and Algebra, 1580–1700." *Isis* 88 (1997): 674–92.

Pocock, J. G. A. *The Machiavellian Moment*. Princeton, N.J.: Princeton University Press, 1975.

Popper, Karl R. *The Logic of Scientific Discovery*. 1935. London: Routledge, 1992.

Pratt, Mary Louise. *Imperial Eyes*. London: Routledge, 1992.

Purchas, Samuel. *Purchas his Pilgrimes*. Vol. 14. London, 1625.

Quinn, David B., ed. *The Hakluyt Handbook*. London: Hakluyt Society Publications, series 2, no. 144, 1974.

———. "Thomas Harriot and the New World." In *Thomas Harriot, Renaissance Scientist*, ed. John W. Shirley, 36–53. Oxford: Clarendon Press, 1983r.

———. "Thomas Harriot and the Virginia Voyages of 1602." *William and Mary Quarterly*, 3d series, 27 (1970): 268–81.

———. *Set Fair for Roanoke*. Chapel Hill: University of North Carolina Press, 1984.

———. *The Roanoke Voyages 1584–1590*. Nendeln, Liechtenstein: Kraus Reprint, 1967.

———. *The Voyages and Colonising Enterprises of Sir Humphrey Gilbert*. London: Hakluyt Society Publications 2, nos. 83–84, 1940.

Quinn, David B., and John W. Shirley. "A Contemporary List of Hariot References." *Renaissance Quarterly* 22, no. 1 (spring 1969): 9–26.

Raleigh, Walter. *The Discovery of Guiana by Sir Walter Ralegh*. Ed. V. T. Harlow. 1595. Reprint, London: Argonaut Press, 1928.

Recorde, Robert. *The Castle of Knowledge*. London, 1556.

———. *The Grounde of Artes*. London, 1542.

———. *The Pathway to Knowledge*. London, 1551.

———. *The Whetstone of Witte*. London, 1557.

Restivo, Sal. *Mathematics in Society and History*. Boston: Kluwer Academic Publishers, 1992.

———. *The Social Relations of Physics, Mysticism, and Mathematics*. Boston: D. Reidel, 1983.

Rheticus, Georg Joachim. "Narratio Prima." In *Three Copernican Treatises*, ed. Edward Rosen, 107–96. New York: Dover, 1959.

Richards, Joan L. *Mathematical Visions: The Pursuit of Geometry in Victorian England*. San Diego, Calif.: Academic Press, 1988.

Ridley, Marke. *A Short Treatise of Magneticall Bodies and Motions*. London, 1613.

Rigaud, Stephen J. *Correspondence of Scientific Men of the Seventeenth Century*. 2 vols. Oxford: Oxford University Press, 1841.

Rigault, Hyppolyte. *Histoire de la querelle des anciens et des modernes*. Paris: L. Hachette, 1856.

Riley-Smith, Jonathan. *The First Crusade and the Idea of Crusading*. Philadelphia: University of Pennsylvania Press, 1987.

———, ed. *The Oxford Illustrated History of the Crusades*. Oxford: Oxford University Press, 1995.

Robinson, A. H. W. *Marine Cartography in Britain*. Leicester: Leicester University Press, 1962.

Roche, John J. "Harriot's 'Regiment of the Sun' and its Background in Sixteenth Century Navigation." *The British Journal for the History of Science* 14 (1981): 245–61.

Rossi, Paolo. *Philosophy, Technology, and the Arts in the Early Modern Era*. Trans. Salvator Attanasio. New York: Harper & Row, 1970.

Rotman, Brian. *Ad Infinitum: The Ghost in Turing's Machine*. Stanford, Calif.: Stanford University Press, 1993.

———. *Signifying Nothing: The Semiotics of Zero*. Stanford, Calif.: Stanford University Press, 1993.

Rukeyser, Muriel. *The Traces of Thomas Hariot*. New York: Random House, 1970.

Scott, J. F. "Oughtred, William." In *The Dictionary of Scientific Biography*, 16 vols., ed. Charles C. Gillispie. New York: Scribners, 1970–80.

Settle, Dionyse. *Laste Voyage into the West and Northwest Regions*. London, 1577.

Shapin, Steven. *A Social History of Truth*. Chicago: University of Chicago Press, 1994.

Shapin, Steven, and Simon Schaffer. *Leviathan and the Air Pump*. Princeton, N.J.: Princeton University Press, 1985.

Sherman, William H. *John Dee: The Politics of Reading and Writing in the English Renaissance*. Amherst: University of Massachusetts Press, 1995.

Shirley, John W. "An Early Determination of Snell's Law." *American Journal of Physics* 19 (1951): 507-8.

———. *Thomas Harriot: A Biography*. Oxford: Clarendon Press, 1983.

———. *Thomas Harriot, Renaissance Scientist*. Oxford: Clarendon Press, 1974.

Skelton, R. A. *Explorers' Maps*. London: Spring Books, 1958.

Smet, Antoine de. "Thomas Blundeville et l'Histoire de la cartographie du XVIe Siecle." *Rivista da Universidade de Coimbra* 27 (1979): 293-301.

Spedding James, Robert Leslie Ellis, and Douglas Denon Heath, eds. *The Works of Francis Bacon*. London: Longmans, 1889.

Spengler, Oswald. *The Decline of the West*. 2 vols. New York: Alfred A. Knopf, 1926-28.

Steele, Robert. *The Earliest English Arithmetics*. Oxford: Oxford University Press, 1922.

Stefansson, Vilhjalmur, and Eloise McCasskill, eds. *The Three Voyages of Martin Frobisher*. 2 vols. London: Argonaut Press, 1938.

Stevens, Henry. *Thomas Harriot and his Associates*. London: Chiswick Press, 1900.

Stevin, Simon. *De Bghinselen der Weeghconst*. Leyden, 1586.

———. *The Principal Works of Simon Stevin*. 6 vols. Ed. D. J. Struik. Amsterdam: C. V. Swets & Zeitlinger, 1955-66.

Tanner, Rosalind, C. H. "Nathaniel Torporley and the Harriot Manuscripts." *Annals of Science* 25 (1969): 339-49.

———. "Thomas Harriot as Mathematician: A Legacy of Hearsay (part I)." *Physis* 9 (1967): 235-47.

———. "Thomas Harriot as Mathematician: A Legacy of Hearsay (part II)." *Physis* 9 (1967): 257-92.

Tapp, John. *The Pathway to Knowledge; Containing the whole Art of Arithmeticke, both in whole numbers and fractions*. London, 1613.

———. *The Seamans Kalendar, or an Ephemerides of the Sun, Moone, and certaine of the most notable fixed Starres*. London, 1602.

Taylor, E. G. R. "Hariot's Instructions for Raleigh's Voyages to Guiana, 1595." *Journal of the Institute of Navigation* 6 (1952): 345-51.

———. "John Dee and the Nautical Triangle." *Journal of the Institute of Navigation* 8 (October 1955): 318-25.

———. *Late Tudor and Early Stuart Geography 1583-1650*. London, 1934.

———. *The Mathematical Practitioners of Tudor and Stuart England*. Cambridge: Cambridge University Press, 1954.

————, ed. *The Original Writings and Correspondence of the Two Richard Hakluyts.* 2 vols. London: Hakluyt Society Publications, series 2, nos. 76–77, 1935.

————. *Tudor Geography 1485–1583.* London: Methuen, 1930.

Tiles, Mary. *Mathematics and the Image of Reason.* London: Routledge, 1991.

Todorov, Tzvetan. *The Conquest of America.* New York: HarperPerennial, 1992.

Torporley, Nathaniel. "Corrector analyticus." In *A Collection of Letters Illustrative of the Progress of Science in England*, ed. J. O. Halliwell, 109–16. London: R. and J. E. Taylor, 1841.

————. *Diclides coelometricae seu valuae astronomicae universales.* London, 1602.

Torricelli, Evangelista. *De infinitis spiralibus.* Ed. Ettore Carruccio. Pisa: Domus Galilaeana, 1955.

————.*Opere di Evangelista Torricelli.* 4 vols. Ed. Gino Loria and Giuseppe Vassura. Faenza: G. Montanari, 1919.

————. *Opera Geometrica.* Florence, 1641.

————. *Opere Scelte di Evangelista Torricelli.* Ed. Lanfranco Belloni. Turin: Unione Tipografico Editrice Torinese, 1975.

Turnbull, David. *Maps Are Territories.* Chicago: University of Chicago Press, 1993.

Tyacke, Sarah, ed. *English Map Making 1500–1650.* London: The British Library, 1983.

Vickers, Nancy. "Diana Described." *Critical Inquiry* 8 (1981): 265–79.

Vitruvius. *On Architecture.* Trans. F. Granger. Cambridge, Mass.: Harvard University Press, Loeb Classical Library, nos. 251, 280, 1970–83.

Wallis, John. *Arithmetica infinitorum.* London, 1656.

————, ed. *Commercium epistolicum de quaestionibus quibusdam mathematicis nuper habitum.* Oxford, 1658.

————. *A Treatise on Algebra, Both Historical and Practical.* London, 1685.

————. *Truth Tried or, Animadversions on a Treatise Published by the Right Honorable Robert Lord Brooke, Entitled the Nature of Truth.* London, 1642.

Wallis, P. J. "Edward Wright." In *The Dictionary of Scientific Biography*, 16 vols., ed. Charles C. Gillispie. New York: Scribners, 1970–80.

Waters, D. W. *The Art of Navigation in England in Elizabethan and Early Stuart Times.* New Haven, Conn.: Yale University Press, 1958.

Watts, Pauline Moffitt. "Prophecy and Discovery: On the Spiritual Origins of Christopher Columbus's 'Enterprise of the Indies.'" *American Historical Review* 90 (1985): 73–102.

Webster, Charles. *The Great Instauration: Science, Medicine, and Reform, 1626–1660.* London: Duckworth, 1975.

Wittgenstein, Ludwig. *Remarks on the Foundations of Mathematics.* Cambridge, Mass.: Harvard University Press, 1967.

Wood, Dennis. *The Power of Maps.* New York: The Guilford Press, 1992.

Wright, Edward. *Certaine Errors in Navigation, Arising either of the Ordinarie*

Erroneous Making or Using of the Sea Chart, Compasse, Crosse Staffe, and Tables of Declination of the Sunne, and Fixed Starres Detected and Corrected. London, 1599.

———. "A Cruizing Voyage to the Azores in 1589 by the Earl of Cumberland." In *A General History and Collection of Voyages and Travels,* vol. 7, ed. Robert Kerr, 375–96. Edinburgh: W. Blackwood, 1824.

Yates, Frances. *Astraea: The Imperial Theme in the Sixteenth Century.* London: Routledge, 1975.

Zelinsky, Wilbur. "The First and Last Frontier of Communication: The Map as Mystery." Special Libraries Association, *Geography and Map Division Bulletin* no. 94 (December 1973): 2–8.

INDEX

Italicized page numbers refer to map illustrations.